ENTOMOLOGY AND PEST MANAGEMENT

ENTOMOLOGY AND PEST MANAGEMENT

Harry Cedric

Gina Gunn

Mathew Wade

Kruger Brentt
Publishers

2023

Kruger Brentt Publishers UK. LTD.
Company Number 9728962

Regd. Office: 68 St Margarets Road, Edgware, Middlesex HA8 9UU

© 2023 AUTHORS
ISBN: 9781787150430

For information on all our publications visit our website at http://krugerbrentt.com/

PREFACE

Entomology is the study of insects and their relationship to humans, the environment, and other organisms. Entomologists make great contributions to such diverse fields as agriculture, chemistry, biology, human/animal health, molecular science, criminology, and forensics. The study of insects serves as the basis for developments in biological and chemical pest control, food and fiber production and storage, pharmaceuticals epidemiology, biological diversity, and a variety of other fields of science. Professional entomologists contribute to the betterment of humankind by detecting the role of insects in the spread of disease and discovering ways of protecting food and fiber crops, and livestock from being damaged. They study the way beneficial insects contribute to the well-being of humans, animals, and plants. Amateur entomologists are interested in insects because of the beauty and diversity of these creatures.

During ancient times, humans had to live with and tolerate the ravages of insects and other pests, but gradually learned to improve their condition through trial-and-error experiences. Over the centuries, farmers developed a number of mechanical, cultural, physical and biological control measures to minimize the damage caused by phytophagous insects. Synthetic organic insecticides developed during the mid-twentieth century initially provided spectacular control of these insects and resulted in the abandonment of traditional pest control practices. This was followed by the development of high yielding varieties of important crop plants. The intensive cultivation of these varieties, together with the application of increasing amounts of fertilizers and pesticides, has resulted in a manifold increase in productivity. However, this technology package has also resulted in aggravation of pest problems in agricultural crops. The importance of achieving sustainable food production through the use of eco-friendly sustainable pest management techniques is being realized more and more in the recent past. Hence the increasing problems encountered with insecticide use resulted in the origin of the integrated pest management (IPM) concept.

The present book contains "***Entomolgy and Pest Management***" chapters covering all related disciplines. These chapters include Entomology and Pest Management, Principles of Insect Pests' Management, Insect Physiology, Methods of Beekeeping, Methods of Bee keeping, Economic Importance of Insects, Insect Pests of Cereal Crops and their Management, Insect Pests of Horticultural Crops and its Management, Insect Pests of Stored Grain and their Management. This book explores ecologically sound and innovative techniques in insect pest management in crops. Related terminology is given at the end for ready reference. It will cater to the requirements of the entomologists, and a valuable source of information for scientists and students in agronomy, botany, disease biology, ecology, evolutionary biology, forestry, genetics, horticulture, parasitology, toxicology, and zoology.

We are grateful to all those persons as well as various books, manuals, periodicals, magazines, journals etc. that helped in the preparation of this book. In spite of the best efforts, it is possible that some errors may have occurred into the compilation and editing of the book. Further queries, constructive suggestions and criticisms for the improvement of the book are always welcome and shall be thankfully acknowledged.

Harry Cedric

Gina Gunn

Mathew Wade

CONTENTS

1

INTRODUCTION TO
ENTOMOLOGY AND PEST MANAGEMENT

1.1 BASICS OF ENTOMOLOGY

Insects and mites are among the oldest, most numerous, and most successful creatures on earth. It is estimated that over 100,000 different species live in North America. In the typical backyard, there are probably 1,000 insects at any given time. While insects which cause problems for humans are heard about most often, it is important to note that the vast majority are either beneficial or harmless. Insects pollinate fruits and vegetables, provide food for birds and fish, and produce useful products such as honey, wax, shellac, and silk. In addition, some insects are beneficial because they feed on other insects that are considered pests by humans. Although the number of pest species compared to the total number of insect species is very small (less than 3% of all insects are classified as pests), the troubles for humans wrought by this group reach astonishing proportions. Insects annually destroy millions of dollars' worth of crops, fruits, shade trees and ornamental plants, stored products, household items, and other materials valued by man. They transmit diseases of humans and domestic animals. They attack humans and pets, causing irritation, blood loss, and in some instances, death.

1.2 BASICS OF CLASSIFICATION

Identification of the thousands of species of insects would be impossible if they were not organized around a standard classification system. By grouping organisms based on the degrees of similarity among them, we can arrive at a system of classification. At the highest level of this classification system, organisms are divided into five kingdoms. Insects are placed in the Animal Kingdom. The Animal Kingdom has major divisions known as **phyla.** Several of the phyla which contain agricultural pests are:

⦿ Arthropoda (insects, spiders, crayfish, millipedes)

The abdomen may have 11 or 12 segments, but in most cases they are difficult to distinguish. Some insects have a pair of appendages at the tip of the abdomen. They may be short, as in grasshoppers, termites, and cockroaches; extremely long, as in mayflies; or curved, as in earwigs.

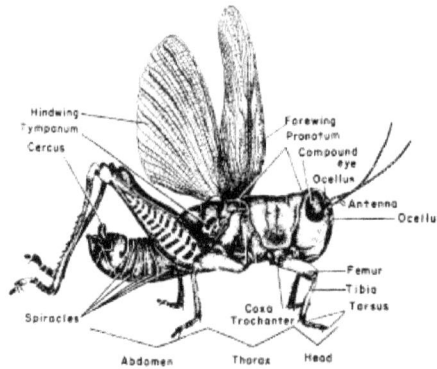

1.3.2 Legs

The most important characteristic of an insect is the presence of three pairs of jointed legs. These are almost always present on adult insects and are generally present in the other stages as well. In addition to walking and jumping, insects often use their legs for digging, grasping, feeling, swimming, carrying loads, building nests, and cleaning parts of the body. The legs of insects vary greatly in size and form and are used in classification.

Leg adaptations of some insects (left to right): jumping (grasshopper), running (beetle), digging (mole cricket), grasping (praying mantis), swimming (diving beetle).

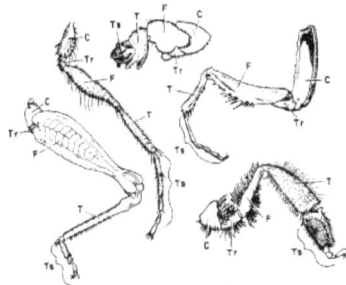

1.3.4 Wings

Venation (the arrangement of veins in wings) is different for each species of insect; thus, it serves as a means of identification. Systems have been devised to designate the venation for descriptive purposes. Wing surfaces are covered with fine hairs or scales, or they may be bare. Note that the names of many insect orders end in "-ptera," which comes from the Greek word meaning "with wings." Thus, each of these names denotes some feature of the wings. Hemiptera means half-winged; Hymenoptera means membrane-winged; Diptera means two-winged; Isoptera means equal wings.

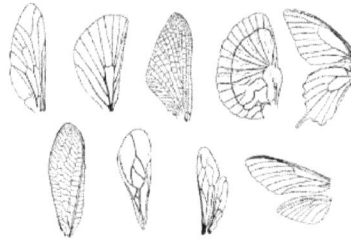

1.3.5 Antennae

The main features of the insect's head are the eyes, antennae, and mouthparts. The antennae are a prominent and distinctive feature of insects. Adult insects have one pair of antennae located on the head usually between or in front of the eyes. Antennae are segmented, vary greatly in form and complexity, and are often referred to as horns or "feelers." They are primarily organs of smell, but serve other functions in some insects.

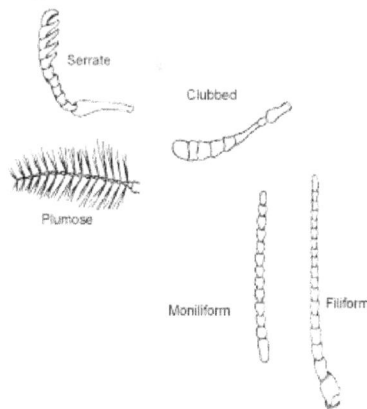

Serrate

Clubbed

Plumose

Moniliform

Filiform

1.3.6 MOUTHPARTS

The most remarkable and complicated structural feature of an insect is the mouth. Great variations exist in form and function of insect mouthparts. And although insect mouthparts differ considerably in appearance, the same basic parts are found in all types. Most insects are divided into two broad categories by the type of mouthparts they possess — those with mouthparts adapted for **chewing** and those with mouthparts adapted for **sucking.**

There are intermediate types of mouthparts: rasping-sucking, as found in thrips; and chewinglapping, as found in honey bees, wasps, and bumble bees. Sucking types are greatly varied. Piercingsucking mouthparts are typical of the Hemiptera (bugs), Homoptera (aphids, scales, mealybugs), blood-sucking lice, fleas, mosquitoes, and the socalled biting flies. In the siphoning types, as seen in butterflies and moths, the mandibles are absent and the labial and maxillary palpi are greatly reduced. Houseflies have sponging mouthparts.

Some types of insect mouthparts: A. Chewinglapping (honey bee); B. Piercing-sucking (plant bug); C. Sponging (housefly); D. Siphoning, coiled (butterfly)

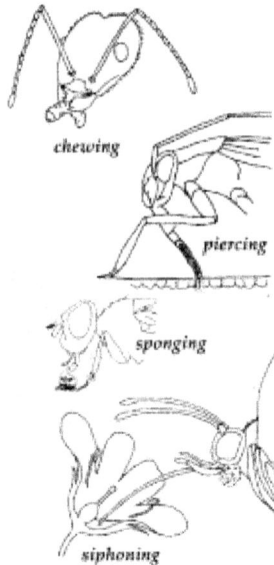

The mouthparts of immature insects tend to be more varied than those of the adults, although nymphs have mouthparts similar to those of the adults. Larval forms generally have the chewing type regardless of the kind possessed by the adults. In some adult insects, the mouthparts are vestigial (no longer used for feeding).

1.4 INSECT DEVELOPMENT — METAMORPHOSIS

In higher animals, the most important development takes place before birth (in the embryonic stage); in insects, it occurs after birth or egg hatch. The immature period of an insect is primarily one of growth, feeding, and storing up food for the pupal and adult stages which follow. Many insects feed very little or not at all during their adult lives.

One of the distinctive features of insects is the phenomenon called metamorphosis. The term is a combination of two Greek words: *meta,* meaning change, and *morphe,* meaning form. It is commonly defined as a marked or abrupt change in form or structure, and refers to all stages of development. Insects undergo one of four types of metamorphosis.

Some insects do not go through a metamorphosis, but rather gradually increase in size while maintaining the same characteristics. Others experience a gradual metamorphosis, going through a nymph stage.

In the case of gradual metamorphosis, the stages are: **egg, nymph, and adult**. In some insects, fertilization of the egg by sperm is not necessary for reproduction. This type of reproduction is known as parthenogenesis. Aphids are notable examples of insects that can reproduce by parthenogenesis.

Insects that undergo complete metamorphosis go through the following stages: **egg, larva, pupa, and adult.**

The immature insect sheds its outer skeleton (molts) at various stages of growth, since it outgrows the hard covering or cuticle more than once. Most insects do not grow gradually as many other animals do. They grow by stages. When their skeleton gets too tight, it splits open and the insect crawls out, protected by a new and larger skeleton that has formed underneath the old one. The stage of life between each molt is called an **instar**. Following each molt, the insect increases its feeding. The number of instars, or frequency of molts, varies considerably between species and to some extent with food supply, temperature, and moisture.

The pupal stage is one of profound change. It is a period of transformation from larva to adult. Many tissues and structures, such as prolegs, are completely broken down and true legs, antennae, wings, and other structures of the adult are formed.

The adult insect does not grow in the usual sense. The adult period is primarily one of reproduction and is sometimes of short duration. Their food is often entirely different from that of the larval stage.

1.5 IDENTIFYING INSECTS

Most home gardeners can classify an insect by the common name of its order, identifying it as a beetle, wasp, or butterfly. The ability to classify an insect to the order level gives the gardener access to much valuable information. This information would include the type of mouthparts the insect has (this tells us how it feeds and gives clues towards methods of control), its life cycle (and proper timing for best control), and type of habitation.

Specific Insect Orders. For your reference, the insect orders have been divided into three sections: those containing insects important to the gardener; those containing insects of lesser importance to the gardener; and common "non-insect" pests in New England. The orders containing insects of importance to home gardeners will be considered in detail.

1.6. INSECT ORDERS IMPORTANT TO THE GARDENER:

Coleoptera - Beetles, Weevils

- ⊙ Adults have hardened, horny, outer skeleton
- ⊙ Adults have two pairs of wings, the outer pair hardened and the inner pair membranous
- ⊙ Chewing mouthparts
- ⊙ Adults usually have noticeable antennae

- Larvae with head capsule, three pairs of legs on the thorax, no legs on the abdomen. Weevil larvae lack legs.
- Complete metamorphosis

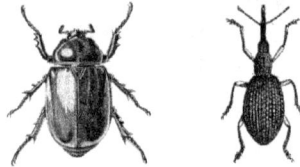

Dermaptera – Earwigs

- Adults are moderate-sized insects
- Chewing mouthparts
- Gradual metamorphosis
- Elongate, flattened insects with strong, movable forceps on the abdomen
- Short, hardened outer wings; folded, membranous, "ear-shaped" inner wings
- Adults and nymphs similar in appearance

Diptera - Flies, Mosquitoes, Gnats, Midges

- Adults have only one pair of wings, are rather soft-bodied, and are often hairy
- Adults have sponging (housefly) or piercing (mosquito) mouthparts
- Larvae may have mouth hooks or chewing mouthparts
- Most larvae are legless
- Larvae of advanced forms, housefly and relatives, have no head capsule, possess mouth hooks, and are called maggots; lower forms, such as mosquito larvae and relatives, have a head capsule
- Complete metamorphosis

Hemiptera - Stink Bug, Plant Bug, Squash Bug, Boxelder Bug

- ⊙ Have gradual metamorphosis; stages are egg, nymph, adult
- ⊙ Have two pairs of wings; second pair is membranous, the first pair are "half-wings" -- membranous with thickening on basal half
- ⊙ Adults and nymphs usually resemble one another
- ⊙ Have piercing-sucking mouthparts
- ⊙ Adults and nymphs are both damaging stages

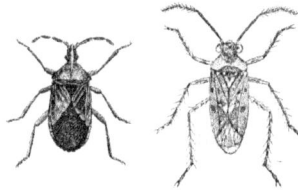

Homoptera - Scale Insects, Mealybugs, Whiteflies, Aphids, Cicadas, Leafhoppers.

- ⊙ Generally small, soft bodied-insects; cicadas may be large and hard-bodied
- ⊙ Winged and unwinged forms
- ⊙ All stages have sucking mouthparts
- ⊙ Have gradual metamorphosis
- ⊙ Many are carriers of plant pathogens

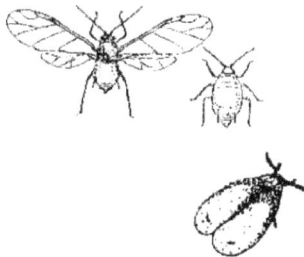

Hymenoptera - Bees, Ants, Wasps, Sawflies, Horntail

- Adults have two pairs of membranous wings
- Larvae have no legs (wasps, bees, ants) or three pairs of legs on thorax and more than four pair of legs on abdomen (some sawflies)
- Generally have chewing mouthparts
- Rather soft-bodied or slightly hard-bodied adults
- Complete metamorphosis

Lepidoptera - Butterflies, Moths

- Adults are soft-bodied, with four well-developed membranous wings covered with small scales
- Larvae have chewing mouthparts
- Adult mouthparts are a coiled, sucking tube; feed on nectar
- Larvae are caterpillars; worm-like, variable in color, voracious feeders
- Larvae generally have legs on the abdomen as well as the thorax
- Complete metamorphosis

Neuroptera - Lacewings, Antlions, Snakeflies, Mantispids, Dobsonfly, Dustywing, Alderfly

- Insect predators, many are aquatic
- Two pairs of membranous wings

- Chewing mouthparts
- Complete metamorphosis

Orthoptera - Grasshopper, Cricket, Praying Mantid

- Adults are moderate to large, often rather hardbodied
- Gradual metamorphosis
- Adults usually have two pairs of wings. Forewings are elongated, narrow, and hardened; hindwings are membranous with extensive folded area
- Chewing mouthparts; both adults and nymphs are damaging
- Hind legs of many forms are enlarged for jumping
- Immature stages are called nymphs and resemble adults, but are wingless

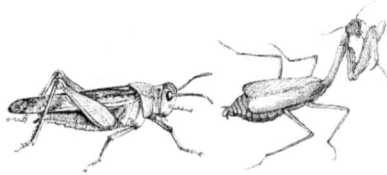

Thysanoptera – Thrips

- Adults are small, soft-bodied insects
- Mouthparts are rasping-sucking
- Varied metamorphosis (a mixture of complete and gradual)
- Found on flowers or leaves of plants
- Wings in two pairs, slender, feather-like, with fringed hairs

1.7 INSECT ORDERS OF LESSER IMPORTANCE TO THE GARDENER:

Order	Examples
Anoplura	Sucking lice
Collembola	Springtails
Diplura	No common examples
Ephemeroptera	Mayflies
Embioptera	Webspinners
Isoptera	Termites
Mallophaga	Chewing lice
Mecoptera	scorpionflies
Odonata	Dragnflies and damselflies
Plecoptera	Stoneflies
Protura	No common examples
Psocoptera	Booklice, barklice
Siphonaptera	Fleas
Strepsiptera	No coomon examples
Thysanura	Silverfish and bristletails
Trichoptera	Caddisflies
Zoraptera	No coomon examples

Common "Non-Insect" Pests

Arachnida - Spiders, Spider Mites, Ticks

a. Spider mites: tiny, soft-bodied animals with two body regions, thick waists, four pairs of legs, no antennae.

Common species:

- Two-spotted mites and near relatives - two spots on the back, may be clear, green, orange, or reddish; usually hard to see without a magnifying glass.

- European red mite - carmine red with white spines.

- Clover mites - brown or gray, flat, very long front legs.

    b. Spiders: resemble mites except that most are larger, and the two body regions are more clearly distinct from one another (thin waist). Most spiders are beneficial predators.

    c. Ticks: resemble large mites and are important agriculturally and medically in that they are parasites of animals and humans.

Diplopoda – Millipedes

These are elongated invertebrates with two visible body regions: the head and body.

They generally have a round cross section, and all but the first four or five body segments possess two pairs of legs. Millipedes are generally inoffensive creatures that feed on fungus and decaying plant material, but at times, they can be fairly destructive to vegetables or plants in greenhouses.

Chilopoda – Centipedes

Centipedes strongly resemble millipedes, except that they have longer antennae, a flat crosssection, and only one pair of legs on each body segment. They are beneficial predators of other arthropods.

Crustacea - Sowbugs, Pillbugs

These are oval with a hard, convex, outer shell made up of a number of plates. Sowbugs are highly dependent on moisture. Generally, they feed on decaying plant material, but they will sometimes attack young plants.

1.8 TYPES OF INSECT INJURY

1.8.1 Injury by Chewing Insects

Insects take their food in a variety of ways. One method is by chewing off the external parts of a plant. Such insects are called chewing insects. It is easy to see examples of this injury. Perhaps the best way to gain an idea of the prevalence of this type of insect damage is to try to find leaves of plants which have no sign of injury from chewing insects. Cabbageworms, armyworms, grasshoppers, Colorado potato beetles, and fall webworms are common examples of insects that cause injury by chewing

1.8.2 Injury by Piercing-Sucking Insects

A second important way insects feed on growing plants is by piercing the epidermis (skin) and sucking sap from plant cells. In this case, only internal liquid portions of the plant are swallowed, even though the insect feeds externally on the plant. These insects have a slender, sharp, pointed portion of the mouthparts which are thrust into the plant and through which sap is sucked. This results in a very different, but nonetheless severe injury. The hole made in this way is so small that it cannot be seen with the unaided eye, but the withdrawal of the sap results in minute spotting of white, brown, or red on leaves, fruits, or twigs; curling leaves; deformed fruit; or general wilting, browning, and dying of the entire plant. Aphids, scale insects, squash bugs, leafhoppers, and plant bugs are wellknown examples of piercing-sucking insects

1.8.3 Injury by Internal Feeders

Many insects feed within plant tissues during part or all of their destructive stages. They gain entrance to plants either in the egg stage, when their mothers deposit eggs into the plant tissue, or after they hatch from the eggs, by eating their way into the plant. In either case, the hole of entry is almost always minute and often invisible. A large hole in a fruit, seed, nut, twig, or trunk generally indicates where the insect has come out, not where it entered.

The chief groups of internal feeders are indicated by their common group names: borers in wood or pith; worms or weevils in fruits, nuts, or seeds; leaf miners; and gall insects. Each group, except the third, contains some of the foremost insect pests of the world. Nearly all of the internal feeding insects live inside the plant during only part of their lives, and emerge usually as adults. Control measures are most effective when aimed at emerging adults or the immature stages prior to entrance into the plant.

Leaf miners are small enough to find comfortable quarters and an abundance of food between the upper and lower epidermis of a leaf.

1.8.4 Injury by Subterranean Insects

Almost as secure from human attack as the internal feeders are those insects that attack plants below the surface of the soil. These include chewers, sap suckers, root borers, and gall insects. The attacks differ from the above-ground forms only in their position with reference to the soil surface. Some subterranean insects spend their entire life cycle below ground. For example, the woolly apple aphid, as both nymph and adult, sucks sap from roots of apple trees causing the development of tumors and subsequent decay of the tree's roots. In other subterranean insects, there is at least one life stage that has not taken up subterranean habit. Examples include wireworms, root maggots, pillbugs, strawberry root weevils, and grape and corn rootworms. The larvae are root feeders, while the adults live above ground.

1.8.5 Injury by Laying Eggs

Probably 95% of insect injury to plants is caused by feeding in the various ways just described. In addition, insects may damage plants by laying eggs in critical plant tissues. The periodical cicada deposits eggs in one-year-old growth of fruit and forest trees, splitting the wood so severely that the entire twig often dies. As soon as the young hatch, they desert the twigs and injure the plant no further.

Gall insects sting plants and cause them to produce a structure of deformed tissue. The insect then finds shelter and abundant food inside this plant growth. It is not known exactly what makes the plants form these elaborate structures when attacked by the insects. However, it is clear that the growth of the gall is initiated by the oviposition of the adult (laying eggs inside plant tissue), and its continued development results from secretions of the developing larva. The same species of insect on different plants causes galls that are similar, while several species of insects attacking the same plant cause galls that are greatly different in appearance. Although the gall is entirely plant tissue, the insect controls and directs the form and shape it takes as it grows.

1.8.6 Use of Plants for Nest Materials

Besides laying eggs in plants, insects sometimes remove parts of plants for the construction of nests or for provisioning nests. Leaf-cutter bees nip out rather neat, circular pieces of rose and other foliage, which are carried away and fashioned together to form thimble-shaped cells.

1.8.7 Ways in Which Insects Injure Plants

- **Chewing** - Devouring or notching leaves; eating wood, bark, roots, stems, fruit, seeds; mining in leaves. Symptoms: ragged leaves, holes in wood and bark or fruit and seed, serpentine mines or blotches, wilted or dead plants, or presence of "worms."

- **Sucking** - Removing sap and cell contents and injecting toxins into plant. Symptom: usually off-color, misshapen foliage and fruit.

- **Vectors of diseases** - Carrying pathogens from plant to plant, e.g., elm bark beetle Dutch elm disease, various aphids - virus diseases. Symptoms: wilt; dwarf,off-color foliage.

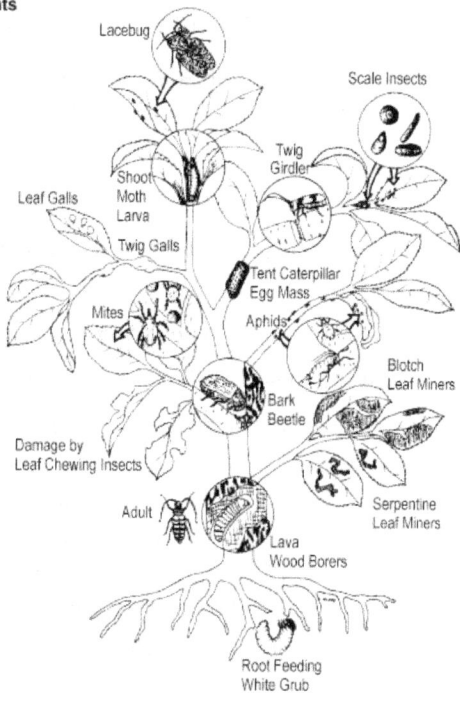

- **Excretions** - Honeydew deposits lead to the growth of sooty mold, and the leaves cannot perform their manufacturing functions. A weakened plant results. Symptoms: sooty black leaves, twigs, branches, and fruit.

- **Gall formation** - galls may form on leaves, twigs, buds, and roots. They disfigure plants, and twig galls often cause serious injury.

- **Oviposition scars** - Scars formed on stems, twigs, bark, or fruit. Symptoms: scarring, splitting, breaking of stems and twigs, misshapen and sometimes infested fruit.

- **Injection of toxic substances** - Symptoms: Scorch, hopper burn.

Examples of insect injury to plants.

1.9 INSECTS AS DISSEMINATORS OF PLANT DISEASES

In 1892, it was discovered that a plant disease (fire blight of fruit trees) was spread by an insect (the honeybee). At present, there is evidence that more than 200 plant diseases are disseminated by insects. The majority of them, about 150, belong to the group known as viruses, 25 or more are due to parasitic fungi, 15 or more are bacterial diseases, and a few are caused by protozoa or mycoplasms.

Insects may spread plant diseases in the following ways:

- by feeding, laying eggs, or boring into plants, they create an entrance point for a disease that is not actually transported by them;
- they carry and disseminate the causative agents of the disease on or in their bodies from one plant to a susceptible surface of another plant;
- they carry pathogens on the outside or inside of their bodies and inject plants hypodermically as they feed;
- the insect may serve as an essential host for some part of the pathogen's life cycle, and the disease could not complete its life cycle without the insect host.

Examples of insect-vectored (insect-carried) plant diseases, their causative agents, and vectors include:

Disease	Vector
Dutch Elm Disease (fungus)	Small Beetle
Fireblight (bacterial)	Pollinating Insects
Tomato spotted wilt (virus)	Thrips
Cucumber Mosaic (virus)	Aphids
X-Disease of Peach (mycoplasm)	Leafhoppers

1.10 BENEFITS AND VALUE OF INSECTS

Insects must be studied carefully to distinguish the beneficial from the harmful. People have often gone to great trouble and expense to destroy insects, only to learn later that the insect destroyed was not only harmless, but was actually saving their crops by eating destructive insects.

Insects are Beneficial to the Gardener in Several Ways:

- Insects aid in the production of fruits, seeds, vegetables, and flowers by pollinating the blossoms. Most common fruits are pollinated by insects. Melons, squash, and many other vegetables require insects to carry their pollen before fruit set. Many ornamental plants, both in the greenhouse and outdoors, are pollinated by insects (chrysanthemums, iris, orchids, and yucca).
- Insects destroy various weeds in the same ways that they injure crop plants.

- Insects improve the physical condition of the soil and promote its fertility by burrowing throughout the surface layer. Also, the dead bodies and droppings of insects serve as fertilizer.
- Insects perform a valuable service as scavengers by devouring the bodies of dead animals and plants and by burying carcasses and dung.

Beneficial Insects

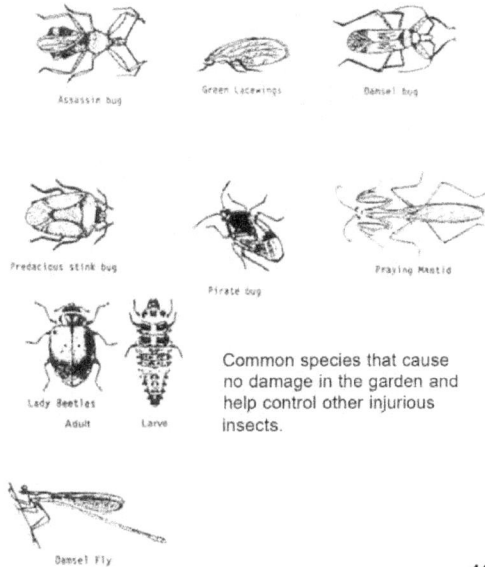

Assassin bug

Green Lacewings

Damsel bug

Predacious stink bug

Pirate bug

Praying Mantid

Lady Beetles
Adult   Larva

Common species that cause no damage in the garden and help control other injurious insects.

Damsel Fly

Many of the benefits from insects enumerated above, although genuine, are insignificant compared with the good that insects do fighting among themselves. There is no doubt that the greatest single factor in keeping plant-feeding insects from overwhelming the rest of the world is that they are fed upon by other insects. Insects that eat other insects are considered in two groups known as predators and parasites.

Predators are insects (or other animals) that catch and devour other creatures (called the prey), usually killing and consuming them in a single meal. The prey is generally smaller and weaker than the predator.

Parasites are forms of living organisms that live on or in the bodies of living organisms (called the hosts) from which they get their food, during at least one stage of their existence. The hosts are usually larger and stronger than the parasites, and are not killed promptly. Some continue to live in close association with the parasite, rather than be killed.

1.11 SOIL PREPARATION

For most fruits and vegetable crops, maintain a slightly acidic soil (around pH 6.5). If in doubt, have a soil analysis done through your local Extension office. The appropriate pH allows vegetable plants to have access to all the necessary soil nutrients and provides a suitable environment for earthworms and microorganisms. Follow recommended fertilizer practices. Supplement chemical fertilizers with organic material or compost to help assure that all trace elements and major nutrients are available. Feed the soil, not just the plants; providing an appropriate environment for all soil life will result in healthy plants which can resist pests and diseases

When using manure and compost, be sure they are worked into the soil. Otherwise, millipedes, white grubs, and other pests may be encouraged. If these insects become a problem, you may be using too much; consider other means of adding organic matter, such as cover-cropping or mulching.

When diseased plant material is added to compost to be used on the garden, delay using the compost until all material has decayed beyond recognition. Compost piles must be hot (160 degrees F.) to kill disease organisms, insect eggs, and weed seeds.

Till the soil in the fall to expose those stages of pests which live near the surface of the soil to natural enemies and weather, and to destroy insects in crop residues. If you do not till in the fall, do so early enough in the spring to give remaining vegetation time to degrade before planting time.

Till the soil in the fall to expose those stages of pests which live near the surface of the soil to natural enemies and weather, and to destroy insects in crop residues. If you do not till in the fall, do so early enough in the spring to give remaining vegetation time to degrade before planting time.

1.12 PLANT SELECTION

Plant crops that are suited to the soil and climate of your area. If you do plant vegetables or fruits that are not normally grown in your area, do your best to provide necessary conditions. For example, watermelons prefer a light, warm, well-drained soil; don't try to plant in heavy clay without first adding copious amounts of compost or other soil-lightening material, and allow the soil to warm up before seeding or setting plants out.

Use disease-free, insect-free, certified seed if available. Select disease/insect-resistant or tolerant species and varieties. Resistance in plants is likely to be interpreted as meaning immune to damage. In reality, it distinguishes plant varieties that exhibit less insect or disease damage when compared with other varieties under similar circumstances. Some varieties may not taste as good to the pest. Some may possess certain physical or chemical properties which repel or discourage insect feeding or egg laying. Some may be able to support insect populations with no appreciable damage or alteration in quality or yield.

Select plants that are sturdy and have well-developed root systems. Diseases and insects in young seedlings may start in greenhouses or plant beds and cause heavy losses in the garden. Buy plants from a reputable grower who can assure you that they are disease/insect-free, or grow your own from seed.

Avoid accepting plants from friends if there is any chance of also getting free insects or diseases!

1.13 CULTURAL PRACTICES

The most effective and most important of all practices is to observe what is going on in the garden. Many serious disease or insect problems can be halted or slowed early by the gardener who knows what to look for and regularly visits the garden for the purpose of trouble-shooting.

1.13.1 Rotation.

Do not grow the same kind of produce in the same place each year. Use related crops in one site only once every three or four years. Some related crops are as follows: (a) chives, garlic, leeks, onions, shallots; (b) beets, Swiss chard, spinach; (c) cabbage, cauliflower, kale, Brussels sprouts, broccoli, kohlrabi, turnips, rutabagas, Chinese cabbage; (d) peas, broad beans, snap beans; (e) carrots, parsley, celery, celeriac, parsnips; (f) potatoes, eggplant, tomatoes, peppers; (g) pumpkins, squash, watermelons, cucumbers, muskmelons; (h) endive, salsify, lettuce.

1.13.2 Interplantings.

Avoid placing all plants of one kind together; alternate groups of different plants within rows or patches. If an insect lays eggs or otherwise attacks a specific species, the presence of other species in the area can interrupt progress of the attack by diluting the odor of the preferred plants. This can also slow the spread of diseases and pests, giving the gardener more time to deal with them.

1.13.3 Interplantings.

Avoid placing all plants of one kind together; alternate groups of different plants within rows or patches. If an insect lays eggs or otherwise attacks a specific species, the presence of other species in the area can interrupt progress of the attack by diluting the odor of the preferred plants. This can also slow the spread of diseases and pests, giving the gardener more time to deal with them.

1.13.4 Watering.

Water in the morning, so plants have time to dry before the cool evening when fungus infection is most likely. Drip irrigation prevents foliage from getting wet at all when watering. For plants susceptible to fungus infections, such as tomatoes, leave extra space between plants to allow good air flow and orient rows so that prevailing winds will help foliage dry quickly after a rain or watering. While this may reduce the number of plants per square foot, yields may still be higher due to reduced disease problems. To prevent spreading diseases, stay out of the garden when the plants are wet with rain or dew.

1.13.5 Time Planting.

Time plantings in such a way that the majority of the crop will avoid the peak of insect infestations. For example, carrot rust fly problems can be avoided by delaying planting until June 1 and harvesting by late August. Keep a record of the dates insect problems occur. Also, by planting warm-weather crops after the soil has warmed, problems with seed and root rots will be avoided, and growth will be more vigorous.

1.13.6 Sanitation

Do not use tobacco products such as cigarettes or cigars when working in the garden. Tomato, pepper, and eggplant are susceptible to a mosaic virus disease which is common in tobacco and may be spread by your hands. Remove infected leaves from diseased plants as soon as you observe them. Dispose of severely diseased plants before they contaminate others. Clean up crop refuse as soon as harvesting is finished. Old sacks, baskets, decaying vegetables, and other rubbish which may harbor insects and diseases should be kept out of the garden.

1.13.7 Staking plants

Staking or planting them in wire cages prevents the fruit from touching the soil. This also helps prevent fruit rots. Caging helps reduce sun scald often seen in staked tomatoes, since caged plants do not require as much pruning, leaving a heavier foliage cover. Boards or a light, open mulch such as straw, placed beneath melons, will prevent rotting.

1.13.8 Avoid Injury to plants

Cuts, bruises, cracks, and insect damage are often sites for infection by disease-causing organisms. In cases where fruit is difficult to remove, such as with cucumbers and watermelons, cut them instead of pulling them off the plant. If you cultivate your garden, avoid cutting into the plant roots.

1.13.9 Mulching

Use a mulch to reduce soil splash, which brings soil-borne pathogens into contact with lower leaves.

1.14 WEED CONTROL

Control weeds and grass. They often harbor pests and compete for nutrients and water. They provide an alternate source of food and can be responsible for pest build-up. They provide cover for cutworms and slugs.

1.14.1 Mechanical Controls

Handpicking. Inspect plants for egg clusters, bean beetles, caterpillars, and other insects as often as possible. Handpick as many as possible. If you don't like squashing the pests, knock the insects and egg clusters into a coffee can or quart jar with a small amount of water, and then pour boiling water over them. Kerosene is often recommended, but poses a disposal problem once you have finished; besides, water is cheaper.

Traps. Use appropriate insect traps to reduce certain insect populations. A simple, effective Japanese beetle trap can be made from two milk jugs or a single milk jug and a plastic bag. The bait used to attract the beetles is available at most farm and garden supply centers. Place traps away from desirable plants. Most scent-based insect traps are used for monitoring populations, not for control of pests.

Light traps, particularly blacklight or blue light traps (special bulbs that emit a higher proportion of ultraviolet light that is highly attractive to nocturnal insects), are good insect-monitoring tools, but provide little or no protection for the garden. While they usually capture a tremendous number of insects, a close examination of light-trap collections shows that they attract both beneficial and harmful insects that would ordinarily not be found in that area. Those insects attracted but not captured remain in the area, and the destructive ones may cause damage later. Also, some wingless species as well as those species only active during the day (diurnal, as opposed to nocturnal) are not caught in these traps. Consequently, the use of a light trap in protecting the home garden is generally of no benefit and, in some instances, detrimental.

Upturned flower pots, boards, newspaper, etc. will trap earwigs, sowbugs, and slugs; collect them every morning, and feed them to pet frogs, toads, turtles, and fish, or destroy them with boiling water. Indoors, white flies can be caught with yellow sticky traps, made with boards painted yellow and lightly coated with oil or grease. There are also commercial sticky traps available through some catalogs.

Barriers. Aluminum foil and other reflective mulch has been shown to repel aphids. However, the environmental impact and energy consumption involved in making aluminum foil deserves consideration. Spread crushed eggshells or hydrated lime around plants to discourage slugs. While heavy mulch is good for weed control, it gives slugs a place to hide.

Exclusion. Various materials can be used to physically block or repel insects and keep them from damaging plants. Place wood ash, cardboard tubes, or orange juice cans around seedlings to keep cutworms away from plant stems. Use paper bags over ears of corn to keep birds and insects out; do not cover until pollination is complete. Net-covered cages over young seedlings will help prevent insect, bird, and rabbit damage. Cheesecloth screens for cold frames and hot beds will prevent insect egglaying; sticky barriers on the trunks of trees and woody shrubs will prevent damage by crawling insects. Floating row covers of spun bonded polyethylene are a little more expensive, but their effectiveness in excluding insects is proven by the number of commercial growers that use them, particularly on cole crops and strawberries. Remember that such materials can exclude pollinating insects.

1.14.2 Biological Controls

Predators, Parasites, and Pathogens. The garden and its surrounding environment are alive with many beneficial organisms that are present naturally; however, they may not be numerous enough to control a pest before damage is done. Actually, parasites and predators (usually other types of insects) are most effective when pest populations have stabilized or are relatively low. Their influence on increasing pest populations is usually minimal since any increase in parasite and predator numbers depends on an even greater increase in pest numbers. Disease pathogens, however, seem to be most effective when pest populations are large.

Take advantage of the biological control already taking place in your garden by encouraging natural predators, such as preying mantids, ladybugs, lacewings, ground beetles, and others. Purchased natural predators are often effective for only a short period, however, since they tend not to remain in the place where they are put. Research the likes and dislikes of these helpers as to foods, habitat, etc. Provide these conditions where possible; some beneficial insect suppliers now offer a formulation for feeding/attracting the beneficials to keep them in the garden longer.

Learn to recognize the eggs and larvae of the beneficial insects, and avoid harming them. You can often find preying mantid egg cases in weedy lots; just bring the twig with the cluster into the garden and set it in a place where it will not be disturbed. Spiders, toads, and dragonflies are also beneficial, and should not be a source of fright to the gardener; in most cases, they are harmless to people.

Learn to recognize parasites and their egg cases; for example, the tomato hornworm is often seen with a number of white cocoons, a little larger than a grain of rice, on its back. These were produced by parasitic wasps. The hornworm will die and more wasps will emerge. You may wish to leave such parasitized caterpillars alone, rather than killing the

1.15 PESTICIDES

1.15.1 Non-synthetic Pesticides

Botanicals. Natural pesticidal products are available as an alternative to synthetic chemical formulations. Some of the botanical pesticides are toxic to fish and other cold-blooded creatures and should be treated with care. Safety clothing should be worn when spraying these, because some may be more toxic than synthetics. Botanical insecticides break down readily in soil and are not stored in plant or animal tissue. Often their effects are not as long-lasting as those of synthetic pesticides.

Insecticide	Use Against
Pyrethrum	Aphids, leafhoppers, spider mites, cabbageworms.
Rotenone	Spittlebugs, aphids, potato beetles, chinch bugs, spider mites, carpenter ants.
Ryania	Coding moths, Japanese beetles, squash bugs, potato aphids, onion thrips, corn earworms.
Sabadilla	Grasshoppers, codling moths, moths, armyworms, aphids, cabbage loopers, blister beetles.

Some of these products may be very difficult to find, expensive, and may not be registered for use in New England.

In addition to botanical insecticides, some biological products can help in the battle against insects. ***Bacillus thuringiensis*** is an effective product commonly used against caterpillars; B.T., as it is known, is a bacterium that gives the larvae a disease, and is most effective on young larvae. Presently, there is research underway to develop strains that work against other types of insect larvae. Several formulations are available to the gardener under different trade names to provide effective control of several caterpillars without harming humans and domestic animals. More than 400 insect species are known to be affected by this important insect pathogen. Bacillus thuringiensis is quite slow in its action. For example, caterpillars that consume some of the spores will stop eating within 2 hours, but may continue to live and move around until they die, which may be as long as 72 hours. When this occurs, the untrained gardener may assume the material was ineffective because of the continued pest activity and impatiently apply a chemical pesticide. ***B.t. kurstaki*** is effective on caterpillars. ***B.t. israelensis*** is used for larvae of mosquitoes, black flies and fungus gnats. ***B.t. san diego*** is used for Colorado potato beetle larvae. ***B.t. bui bui*** may soon be available for Japanese beetle control.

Nosema locustae is a disease organism which shows some promise for controlling grasshoppers. There are claims that this parasite may be effective for up to five years after initial application. In some areas, this parasite is available commercially under different trade names. It is still too early to make extensive claims about its effectiveness in home gardens.

Enlist the aid of birds. In rural areas, chickens, guineas, and other domestic fowl can be released in unused areas of the garden to eat grubs and insects. Wild birds will also help, but they aren't as controllable. Provide appropriate conditions (i.e., shelter, nesting material, water) to encourage insect-eating birds.

Soaps. Commercial insecticidal soap (a special formulation of fatty acids) has been proven effective against aphids, leafhoppers, mealybugs, mites, pear psylla, thrips, and whiteflies. Homemade soap sprays also work to some extent: use three tablespoons of soap flakes (not detergent) per gallon of water and spray on plants until dripping. Repellent sprays, such as garlic sprays and bug sprays (made from a puree of bugs), have been found useful by some gardeners, but their effectiveness is questionable. Some researchers believe that bug sprays may work if a disease is present in the insect, which is spread through the spray to other insects. Be careful! Homemade soap sprays can injure some plants.

1.15.2 Synthetic Pesticides

Synthetic pesticides, by their simplest definition, are those pesticides made by humans in chemical laboratories or factories. Examples of these include malathion, diazinon, and sevin. The real surge of development of synthetic pesticides began in World War II with the discovery of DDT.

Insects constitute one class of the phylum Arthropoda, and yet they are one of the largest groups in the animal kingdom. The insect world is made up of individuals that vary greatly in size, color, and shape. Although most insects are harmless or even beneficial to humans, the few that cause damage have tremendous impact. Harmful species can usually be recognized with some basic knowledge of their host, habits, life cycle, and the type of damage they inflict. Feeding damage varies due to the type of mouthparts an insect possesses. Harmful insects can be controlled in many ways without resorting to the use of pesticides. Good cultural practices and proper selection of plant varieties, coupled with mechanical and biological controls, will control insect populations. Use insecticides judiciously, wisely, and safely. Read and follow label directions carefully when applying any pesticide.

2

PRINCIPLES OF
INSECT PESTS MANAGEMENT

2.1. INTRODUCTION

Every organism interacts with biotic and abiotic components of ecosystem and struggle for its better survival and existence in nature. Different types of interactions exist between pests and other components of ecosystem, especially human, plants and animals. These interactions can create issues of competition for food and space; endemic or epidemic outbreak of diseases/ nuisance, damage to properties and injury to both plants and animals. Pest incidence as well as its population dynamics and control are regulated in such interactive agroecosystem by many forces and factors. The major of these are forces of destruction like environmental resistance, forces of creations like biotic potential and characteristics or components of an agroecosystem. A comprehensive knowledge of destructive forces including abiotic stresses (adverse environmental conditions/density independent factors), biotic stresses (density dependent factors like predators, parasites, pathogens, competitors etc.) and biotic potential (reproductive potential, survival potential, nutritive potential and protective potential) (Fig. 2.1) help us decide whether the time is to adopt "Do-Nothing Strategy", "Reduce-Number Strategy", "Reduce-Crop-Susceptibility Strategy" or "Integrated-Strategy" to manage the indeginous and exotic emerging pest problems (Pedigo and Rice 2009; Schowalter 2011).

Pest management is a two-strand approach which mainly relys on the knowledge of the strategy, pest biology and pest ecology in agroecosystem (Fig. 2.1). The selection of appropriate pest control technology as well as its effective and efficient application mainly depends upon a comprehensive knowledge about it. The biological and ecological knowledge of pest helps to determine the most appropriate procedure/method (How), timing (when) and place (where) for effective use of any technology and economically effective management of any pest (Buurma 2008). Various aspects of any technology which lead towards its proper and effective application include:

- Nature and type of technology
- Method of its application (aerial, foliar, chemigation, baits, traps etc.)
- Durability of the technology
- Equipment's and accessories needed for its implementation
- Performance limiting factors of technology
- Compatibility of technology with other available management tools
- Specificity of the technology (broad spectrum or target specific))
- Mode of action

The knowledge of various aspects of biology and ecology of pests lay the foundation of an efficient and economical pest control strategy is important for achieving key objectives of pest management. For examples, suck kind of knowledge reduces the threat of crop failure by endemic or epidemic pest outbreak. Such knowledge also strengthens the effectiveness of pest control strategies, reduces operational cost of technique used, enhances productivity and profitability by reducing the amount of inputs and ultimately eliminates or reduces the threats of environmental degradation and hazards of human health. Various aspects of pest's biology that can be helpful in devising efficient pest management strategies include:

- What kind of habitat does the pest prefer? (Darkness, indoor, outdoor, humid, warm, temperate, aquatic, terrestrial etc.)
- What kind of food does the pest prefer?
- What is the total life span of pest?
- What is longevity of incubation period of the pest?
- Where are different life stages found?
- What is the breeding place and season of the pest? What kind of behavior does the pest exhibit in its life?

Integrated application of multiple and highly compatible tactics; reduction in number or effects of pest below defined economic decision levels (EIL and ETL); and conservation of environmental quality are the key characteristics/elements of sustainable pest management. Scientists suggests that some supplementary characteristics/elements of sustainable pest management system and deliberates that a pest management technology/ system should be:

1. Highly target specific i.e., very selective for pest and safe for nontarget organisms ;

2. Comprehensive and conducive for crop productivity (not be phytotoxic and enhance plant-growth and yield);

3. Highly compatible with the key principles of ecology and

4. Tolerant to potential pests but within economically tolerable limit. A comprehensive and practical knowledge of above-mentioned elements guarantees the development of an ecofriendly, economical and efficient, crop production and protection program.

Effective and sustainable insect pest management also depends on economic decision levels which are mendatory for determining the course of action, ensuring sensible pesticide application, reducing unacceptable economic damages, safeguarding the profits of producer and conserving the environmental quality in any pest situation .

This chapter highlights the basic concepts and principles of sustainable insect pest management which are based on well-defined goals, estimates of population size, evaluation and comparison of available management options and monitoring as well as evaluation of practiced management activities/strategies regarding benefits and costs. Defined goals determine better and appropriate available management options in a prevailing situation; population estimates help in determining action threshold and deciding timeframe for initiation necessary actions. Evaluation and comparison of available management options, their effectiveness in stipulated time-period, environmental and social consequences and cost-benefit-ratio can help in ranking out the efficient, economical and ecofriendly options. Monitoring as well as evaluation of practiced management activities/strategies on the basis of benefits and costs is too imperative for improved adaptive management. All the principles of pest management are described.

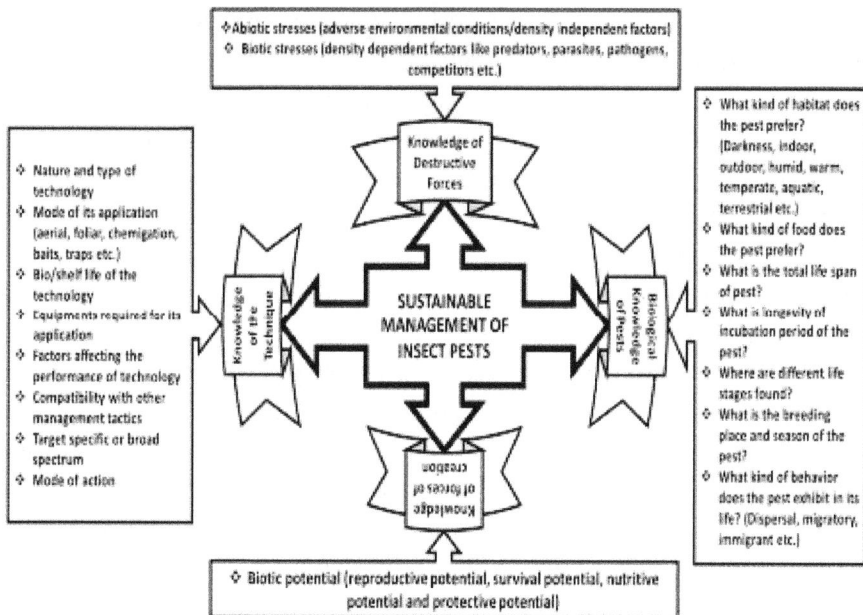

Fig. 1: Four pillars and their associated components which lay down the basis of sustainable management of insect pests.

2.2. GENERAL PRINCIPLES OF PEST MANAGEMENT

The era of conventional crop production and protection has turned to ecofriendly and organic crop production and protection system where natural products based techniques are employed in the agricultural industry and toxic chemicals based techniques are being depleted from the agriculture system. A sustainable agroecosystem system composing of healthier and more productive crops with least utilization of toxic pesticides depends upon a holistic pest management approach (Joshi 2006; Dhaliwal and Koul 2007; Singh 2008) which is based on following basic principles (Fig. 2.3).

2.2.1 Pest avoidance/Exclusion

It is kind of precautionary step which inhibits the entry of any insect pest into any agroecosystem and ensures pest free zone. This principle is based on the utilization of such techniques or practices which exclude and prevent the pest and it is always considered as a foundation step of any IPM program. Pest avoidance or exclusion techniques include hand-picking, screening, bagging, physical beating, banding, trapping, acausting (noise creation), physical barriers, burning, sieving and winnowing and rope dragging, etc., (Dhaliwal et al. 2006).

2.2.2 Hand-Picking

Hand picking is kind of excluding technique which is not practicable for large scale pest management program; however, it can be practiced for small scale pest management program like in lawns, kitchen gardening, small-scale tunnel farming, inside greenhouses. This technique is the most practical way in certain conditions like, when cheap labour is available, insects and their eggs/egg-masses are large and conspicuous, insects are too sluggish, have congregating behavior and are easily accessible to the pickers. Handpicking of slow moving and visible larvae of Pieris brassicae (L.) (Cabbage butterfly) (Lepidoptera: Pieridae), lemon butterfly [Papilio demoleus Linn. (Lepidoptera: Papilionidae)], semiloopers and loopers (Lepidoptera: Noctuidae), cutworms (Lepidoptera: Noctuidae) and red pumkin beetle [Aulacophora foveicollis Lucas (Coleoptera: Chrysomelidae)] and visible eggs/eggmasses of cabbage butterfly, armyworm [Spodoptera (Guenee) and Mythemna (Ochsenheimer,) spp. (Lepidoptera: Noctuidae)], and borers [Pyralid borers, Noctuid borers, Crambid borers etc. (Lepidoptera)] is an easiest, direct and excellent method of controlling them especially when their infestation is restricted to only a few plants. In case of pink bollworm [Pectinophora gossypiella (Saunders) (Lepidoptera: Gelechiidae)] infestation in cotton, the rossetted flowers having pink bollworm larvae inside are picked and destroyed. Collection and destruction of egg masses of top borer [Scirpophaga nivella F. (Lepidoptera: Pyralidae)] in ratoon and seasonal sugarcane (Saccharum officinarum L.) crop reduces its endemic outbreak and losses. For reduction in incidence of gurdaspur borer [Acigona steniellus Hamp. (Lepidoptera: Pyralidae)], area-wide collection and destruction of infested canes harboring its gregarious larvae proves very effective. Deep burying or chemical treatment of fallen fruits reduced the incidence and periodic outbreak of fruit flies in orchards. This conventional and oldest technique helps to collect and destroy the adults before they start laying their eggs, to gather and crash the eggs before they hatch, to pick and kill the larvae/ nymphs before they cause economic losses and ultimately prevents the build-up of pest population and the resulting damage. Clipping and destruction of rosetted flowers infested with P. gossypiella larvae, withered top infested with spotted bollworm [Earias spp. (Fab) (Noctuidae: Lepidoptera)], splayed squares along with sheded fruiting and floral bodies attacked by bollworms, leaves having egg-masses or army of younger larvae of armyworm (Spodoptera spp.) and tobacco caterpillar [Spodoptera litura Fabricius (Lepidoptera: Noctuidae)] can minimize population buildup and extent of cotton damage by these insect pests (Saha and Dhaliwal 2012). Further outbreak and infestation of khappra beetle [Trogoderma granarium Everts (Coleoptera: Dermestidae)] can be reduced if its clustered population is collected and destroyed.

2.2.3. Avoidance/Exclusion by Bagging, Screening and Barriers

Bagging, screening and barriers installation is also considered very useful for protecting the crop and fruits from attack by insect pests as well as for keeping away the insect pests which either act as carrier or vectors of various fatal diseases in animals and man or create nuisance for man. For example, field bags' dragging in the maize (Zea mays L.) or sorghum (Sorghum bicolor L.) field and sugarcane (S. officinarum L.) ratoon crop (April/May) to collect sugarcane pyrilla [Pyrilla perpusilla Walker (Homoptera: Lophopidae)] can reduce the chances of their massive migration from maize/sorghum to sugarcane and population buildup of pyrilla at the initial growth stage of sugarcane ratoon crop. Such type of field bags can also be used for the mass collection of various grasshoppers (Orthoptera: Insecta), bugs (Hemiptera: Insecta), crickets (Orthoptera: Insecta) and other minute, small and large insects harboring vegetation. Wrapping of individual fruits with paper bags, polythene bags, butter-paper bags or net bags protects 95% of these fruits from the infestation of fruit flies. Covering whole small trees with any transparent material can reduce the attack of various insects' pests. Construction of water filled or dust (insecticide) treated drench between wheat (Triticum aestivum L.) field and burseem (Trifolium alexandrium L.) field can reduce the migration of armyworm (Spodoptera spp.) larvae from wheat to burseem fields and minimize their damage on burseem. Similar type of drenches can reduce the migration of bands of locust hoppers [Schistocerca gregaria (Forskål) (Orthoptera: Acrididae)] from the breeding placed to nearest field crops. Application of various types of bands around the tree trunks reduces the upward crawling of various crawling insects and protects their damage to leaves, floral parts and fruits. Sticky bands or funnel type bands installed around the stem of mango tree as a barrier for upward crawling mango mealybug [Drosicha stebbingi Green (Hemiptera: Margarodidae)] stop its nymphs below the bands and protect its damage to inflorescence, tender leaves and mango fruits. Yellow sticky traps are used for the management of aphids (Homoptera: Aphididae) on various crops. Red colored spherical traps are used for the control of fruit flies in orchards. Screening of windows, ventilators and doors of rooms and sheds with fine meshed wire gauze keep the house flies [*Musca domestica* L. (Diptera: Muscidae)], mosquito species [*Culex* spp., *Aedes* spp. (Diptera: Culicidae)] and other insect away from human being and animals which remain protected from the insect borne diseases and nuisance. Heavy irrigation followed by rope dragging on crop cultivated on small scale help to shed away the larger wingless insects or their stages with or without infested plant parts especially flowers, bolls, fruits etc. from the plants into the water and are killed by drowning in water contaminated with insecticides.

2.2.4 Avoidance/Exclusion by Trapping, Shaking, Sieving and Winnowing

Trapping tactic is widely used for the management of insect pest of various economical crops. Various types of traps including light-traps, pheromone-traps, bait-traps, suction traps etc. are used in trapping technique which is being used for monitoring, mass trapping, mating disruption and management of various types of insects. For example, pheromone traps are being practiced successfully for the monitoring, trapping, mating disruption and management of pink bollworm (P. gossypiella Saunders), gypsy moth [Lymantria dispar (L) (Lepidoptera: Lymantriidae)], cotton grey weevil [Anthonomus grandis Boheman (Coleoptera: Curculionidae)], pine beetle [Dendroctonus ponderosae Hopkins (Coleoptera: Curculionidae)], oriental fruit fly [Bactrocera dorsalis (Hendel) (Diptera: Tephritidae)], melon fruit fly [Bactrocera cucurbitae Coquillett (Diptera: Tephritidae)], European chaffer [Rhizotrogus majalis (Raz.) (Coleoptera: Scarabeidae)] . Light traps associated with toxic compound are practiced for the trapping and killing of nocturnal insects pests. Bait-trap consisting of food source as kairomone and odorless insecticide as killing agent is used for attraction, trapping and killing of various insects. GF-120 food-bait is used for the management of fruit flies in orchards and cucurbits crops. Placing chopped turnips (Brassica rapa L.) or potatoes (Solanum tuberosum L.) in form of heap in cutworms (Lepidoptera: Noctuidae) infested crop provide a site of attraction and aggregation for cutworm larvae. From such heaps, the aggregated cutworm larvae can be collected and destroyed. Air-suction trap and tractor mounted light-plus-air suction traps can be employed for attraction and killing of various soft bodies small insects like whiteflies [Bemisia tabaci Genn. (Homoptera: Aleyrodidae)], thrips [Thrips tabaci Lindeman (Thysanoptera: Thripidae)], winged aphids, adults of Dipteran and Lepidopteran leafminers [Liriomyza trifolii (Burgess) (Diptera: Agromyzidae); Phyllonorycter spp. (Lepidoptera: Gracillariidae), Eriocrania spp. (Lepidoptera: Eriocraniidae), citrus leafminer [Phyllocnistis citrella Stainton (Lepidoptera: Phyllocnistinae)], adults of psyllids [(Homoptera: Psylloidea) like Diaphorina citri Kuwayama (Hemiptera: Psyllidae)] etc. The sluggish or immobile insects including mealybugs, aphids, psylids etc. can be separated from the plants/tree-canopy in case of small-scale cultivation, kitchen gardening or landscaping by simple shaking and jarring technique. This technique can also be used against locust and defoliating beetles on small scale. Inside the godowns or any storage structures, various life stages of insect pests of stored grains can be separated, collected and destroyed by sieving and winnowing technique.

2.3. IDENTIFICATION OF PEST AND ITS STATUS: TOOLS AND TECHNIQUES

Identification of pest, its various life stage and its effects is considered the key components of any integrated pest management plan. An accurate identification of pests helps to determine their pest status, population dynamics and effective control measures. However, an accurate identification of prevailing pest species at the spot is a challenging and difficult step too even for any qualified and expert entomologist. Continuous and constant efforts to recognize the prevailing pest species makes correct identification of pests of a particular crop comparatively easier.

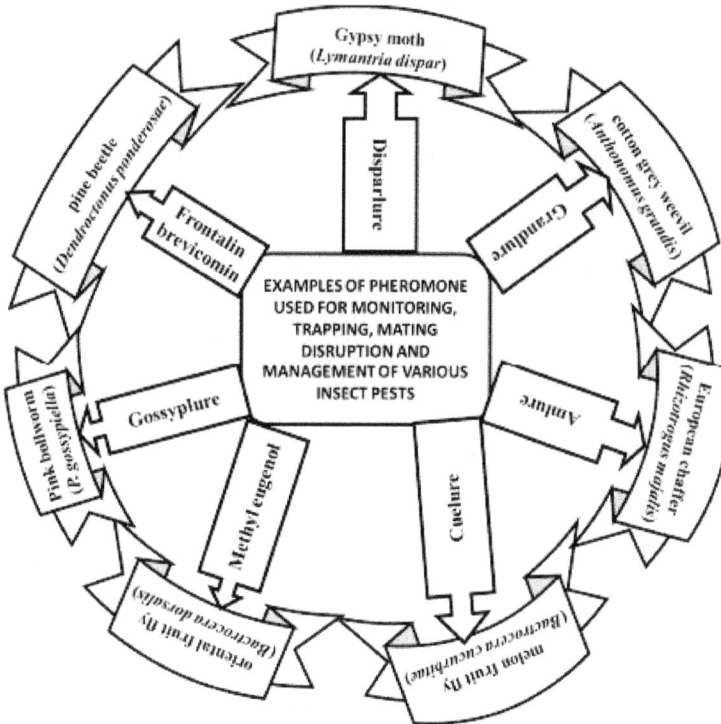

Fig: 2. List of pheromones used for monitoring, trapping, mating disruption and management of various insect pests.

2.3.1 Strategies to Identify Pest Species

Several approaches can be used for an accurate identification of useful and harmful insects. The catches can be compared with pics/images available on the internet websites or in compendium/books. The collections can be sent to expert insect taxonomists or entomologists who can identify insects and explain various questions related to the identified pest. Other strategies include hiring entomologists from local colleges, universities, or pest management companies (pesticide organization) or getting training from these organizations for exact identification of pests. After an accurate identification of the catches, refer ence collection should be maintained for future identification and training of the other stakeholders.

2.3.2 Magnification

For the exact identification of small or minute insects, 15-30 X magnification lens are required. Hand held magnifiers can be used for magnification and identification of small or minute insects.

2.3.3 Identifying insect Debris and Damage

Some insects are very difficult to observe, locate and identify because they mostly remain hiding during daytime and are active for very short period of time or during night. The presence of such insects can be determined based on their debris, damages, remnants, products etc. An intense vigilance is needed. For example, life stages of silverfish [Lepisma saccharina L. (Thysanure: Lepismatidae)], booklice (Psocoptera: Insecta) and other insects are difficult to locate due to their small size, camouflaging color and reclusive habits. The following signs can be used to recognize the commodities under attack of insects:

2.3.4. Insect Remains

Sheded wings of termite, casing of larvae, exuvia of molted insects, dark egg pods of cockroaches, webbing of clothes moths [Tinea pellionella L. (Lepidoptera: Tineidae] are some prominent examples the signs which can be used for identification of insect pests.

2.3.5. Frass

The product of the insect's feeding and digestion is termed as frass. For example, sawdust collected in the excavated holes indicates the damage of powderpost beetle. Mud tunnels on any surface indicate the incidence and damage of termites. The frass color mostly resembles with color of digested food. The presence of frass every time below the infested object is probably the indication ofactive population of infesting insects.

2.3.6. Visible Damage

Feeding damages are visible symptomologies which ensure the incidence of insect pests. The more prominent visible damages include enterrance and exit holes in damaged object or commodity. Presence of clothes moths (T. pellionella L.) may be concentrated on damaged spots on strata. Skinned paper is the indication of occurrence and grazing of silverfish (L. saccharina L.). The activities of wood boring insects can be detected by the exit holes which indicate the damage has already been and adults emerging from the pupae have been escaped.

2.3.7. Smell and Sound

Smell of faeces and decaying byproducts of feeding and digestion by insects also indicate their activities and incidence. Similarly, when insect perform feeding and other life activites, they produce specific sound. Such smell and sounds are detectable and can be perceived and heard, respectively by human. The activities of wood damaging beetles and can be confirmed by hearing special sound produced wood boring insects. The incidence of cockroaches can be defined by their pungent smell. The presence of termites and hollow tunnels/areas under the wood surface can be detected by gently tapping with finger across it and listening the differences in sound produced.

Examples of insects' identification strategies

- ⦿ The adults of termites (Insecta: Isoptera: Kalotermitidae) consume wood or woody materials and create muddy tunnels or hollow areas internally under the wood. Termite activities do not appear in form of flight/exit holes rather sand-like frass and muddy tunnels towards the cellulosic food source (furniture, books, wodden-frames, etc.,) (Peterson et al. 2006).

- ⦿ After hatching from eggs, the larvae of woodboring beetles including powderpost beetles [Xylopsocus capucinus (Fabridius) (Coleoptera: Bostrichidae)] and death watch beetles [Xestobium rufovillosum de Geer (Coleoptera: Anobiidae)], tunnel through the wood and consume it till the emergence of adults through exit holes. A talcum-powder like very fine frass is produced when powderpost beetle consumes wood or woody materials (O'Connor-Marer 2006; Lewis and Seybold 2010).

- ⦿ Dark places with high humidity and starch stocks are preferred by Silverfish (L. saccharina L.) Skimmed papers and other starchy materials indicates its incidence (Nasrin 2016).

- ⦿ Larvae of carpet beetles [Anthrenus flavipes (LeConte) (Coleoptera: Dermestidae)] prefer to graze across the surface of fabric-stuff, plant materials, fur/fleece, and feathers and its damage generally gives a shabby/ragging look to the damage material.

- ⦿ Dermestids include larder beetle [Dermestes lardarius Linnaeus (Coleoptera: Dermestidae)] and hide beetles [D. maculatus DeGeer (Coleoptera: Dermestidae)]. These beetles are very voracious feeders and their larvae voraciuosly devour skins of animals and dried plants or their materials

- ⦿ Identification of case making clothes moths (Tineola pellionella L.) larvae can be easily accomplished by the cases spun around their bodies from the materials they feed on. Similarly, the incidence and damage of webbing clothes moth [T. bisselliella Hummel (Lepidoptera: Tineidae)] larvae can be identified by the silk trail they leave when they move to feed on food.

Both the moth are voracious feeders of woolen-materials, dried woody materials, feathers and fur. Their feeding results in gradual-thinning of the damage materials which can be used as indicators of their attack.

⊙ A tiny pale colored booklice (Psocoptera: Insecta) feed mostly on mold, fungi present in books and on papers. They also feed of dead and decaying insect.

⊙ Bookworms and larvae of cigarette beetle [Lasioderma serricorne F. (Coleoptera: Anobiidae)] and drugstore beetle [Stegobium paniceum (L.) (Coleoptera: Anobiidae)] feed on books, binding materials and dried plant parts. Their feeding results in burrows and holes in the infested materials. These damage patters can be used to identify these beetles when are present or absent in infested materials.

⊙ Cockroaches (Insecta: Blattodea) mostly feed on residues of molds and starchy/protenecious materials deposited on different surfaces including wodden/plastic furnitures, clothings, papers, cardboards etc. Their faeces and drooping cause staining of the mataerials. The female cockroaches adhere their egg pouches to any ptotected objects. Presence of egg-cases and stained materials can be used as indicators of their presence.

2.4. UNDERSTANDING BIOLOGY AND ECOLOGY OF PEST: THEORIES AND PRACTICAL

An efficient, effective and successful management of insect pests is always founded on a comprehensive knowledge of the biology, morphology, internal anatomy, behavior, growth (metamorphosis), life history and ecology of the insect pest. The morphological knowledge of an insect helps to develop an appropriate technology and decide the selection of appropriate insecticide. Chemotropism based techniques involving attractant or repellents have been developed for various insect pests. The development of such techniques depends upon knowledge about chemoreceptors like, gustatory, olfactory, sensory receptors etc. Development and selection of color of light of light-traps depend on the knowledge of structural components and physiology of compound eyes of insects (Dhaliwal and Arora 2003; Pedigo and Rice 2009). The knowledge of structural components and physiology of compound eyes of insect provide information about the type of color which is highly attractive for any insect. For example, yellow sticky traps are used for the control of aphids as aphids are attracted to yellow colour (Saha and Dhaliwal 2012). The knowledge about the types of mouthparts of insect pests helps to decide what type of insecticides should be selected for successful control of the insect pest. For example, if the infesting insect pests have sucking type of mouthparts, then, insecticides with systemic and contact action are the most appropriate. Unlikely, if the infesting insect pests have chewing type of mouthparts then stomach poisons will give effective control. Unawareness of the knowledge of the mouthparts of insect pests leads to wrong selection of insecticides and ineffective management of insect pest in spite of investment of money in form of insecticides application (Saha and Dhaliwal 2012).

Knowledge of internal anatomy and physiology is also very advantageous in devising pest management tactic. For example, so many insecticide molecules with IGR or karomone activity have been discovered and their analogs have been synthesized for commercialization and management of insect pests. These molecules are based on growth hormones, parasitoid's sting-glands peptides and hormones. The knowledge of spiracular respiration can be helpful in controlling the insect pests with fumigation (Hoffmann and Frodsham 1993).

The knowledge of insect metamorphosis and its physiology provides so many useful informations about the week links of insect growth stages and their activity periods and sites which if targeted can ensure for effective management of any pest. Such knowledge can also be useful in synchronizing the timing of application of pest management tactics with week-link or susceptible growth stage of insect; thus, ultimately would be helpful in reducing blind use and application intensity of pesticide on crop. This information would lay the foundation of decision on when, where and how to use available and recommended insecticides or other pest management tactics. Information on the metamorphic stages like, eggs, larvae/nymphs/naiads, pupae and adults of insects comprehend the facts that which stage is notorious, devastating and damaging one and which are not. This information also precises the damaging-stage-specific application of control measures.

Ecological Pest Management (EPM) is one of the key components of IPM. EPM emphasis on targeting many life stages of pest by pest control and minimizing its survival potential with least possible resources/ways (Schowalter 2011). These objectives of EPM can be achieved by following the ecological principles exactly in intimation with nature. It is imperative to devise a production system primarily on the pillars of solid and accurate information, multiple production and protection tactics and ecological principles. Such production system strengthens the individual and combined efficiency of the strategies being used, distribute the plant protection burden across multiple tactices and minimizes the chances of complete crop failure. This system also reduces the blind, chain-repeated and frequent use of one tactice/strategy and resultantly it not only minimizes the rate of resistance development but also lessens its inputs rquirements, reduces operating cost and enhances its productivity and profitability (Pedigo and Rice 2009)

The study of the behavior of insects also laid the foundation of successful control of insect pests. Insect behavioral studies figure out following important facts of their life that can be helpful in controlling them.

2.4.1. Egg laying Behavior

Some insects are endophytic (fruit flies) and some are exophytics (most of the bollworms and borers). Most of the insects deposit exposed eggs while some deposit the covered egg masses. Depending on egg laying behavior, pest management tactic is decided to control insect pest at egg stage.

2.4.2. Behavior of Newly Emerged Young Ones

The young ones of most of the borers just after hatching enter the leaf whorls or stem of the plants, avoid the direct exposure of insecticides and become very difficult to kill with contact insecticides. Similarly, leaf miners just after hatching enter the cortex tissues forming mines and cannot be controlled with contact insecticides. Young ones as well as later instar larvae of cutworm remain hidden in cracks and crevices and insecticides application direct on plants during day time will not yield effective control. Their effective management can be ensured if chemigation of insecticides or bait application is employed.

2.4.3. Feeding Behavior in Young Ones

Feeding habit of insect pest also help in deciding the types of tactics and method of their application for effective management of any insect pest. The insect pest which prefer to feed underside the leaf can be controlled effectively by application of systemic and translamiar insecticides. Similarly, borers (Insecta: Lepidoptera and Coleoptera) exhibit concealed feeding inside the stem which cannot be killed with contact insecticides; rather systemic insecticides will be the most appropriate tactic for the management borers.

2.4.4. Breeding Place

Nipping the evil in the bud for insect pests is possible only if their exact breeding sites are known. It is possible only through comprehensive studies of their biology. The breeding places of mosquitoes are stagnant water and their treatment with larvicides, ovicides or oils help in controlling the breakout of adult population. Cockroaches breed in filthy places which should be targeted with insecticides treatment for their management at bud/root level (breeding places) for terminating their further population buildup and outbreak. For fruit flies, the breeding substrates are dropped fruits which should be collected and destroyed for their population management.

2.5 STUDY OF INSECT ECOLOGY AND IPM

The study of insect ecology provides the conceptual and theoretical framework which offers the practical ground for the application of pest management discipline (Driesche et al. 2008). Interactive effects of insects and other organisms of the agroecosystem influence insect pests management strategy (Schowalter 2011). The solution of insect problems majorly depends on ecological management which is considered as one of the oldest, least expensive and ecologically the most compatible tactics. Ecological studies of insects help in identifying and exploiting the weak links of seasonal life cycle of insects. Such studies also help to explore the food and physical factors which impact insect's life negatively. By manipulating of such factors unfavorable for insect survival, insect pest's outbreak, population buildup and damage impacts can be avoided in an ecofriendly way (Pedigo and Rice 2009). Study of insect ecology also laid the foundation of plant-insect-predators/parasitoids interactions which help to frame out the pest management strategy for any insect pest. According to vegetable system defined by Joe Lewis and Steve Groff (Pattison

2005), a combination of tillage and mulching of vegetables with cover crops conserves sufficient biodiversity for pests. Such integrated practices result in conservation of field and increase beneficial insect populations fourteen times higher than in the conventional fields. A conducive and effective balance between pest and beneficial fauna can be sustained in the crop production system on a farm by maintaining propotionate undisturbed areas on it. The natural enemies (predators and parasites) attacking pests mostly survive, established and conserved in the undisturbed sites like hedges, weedy borders, grassed alleyways, woodlots, grassed waterways, riparian buffers, and small undisturbed areas maintained between rows of major crop. Such undisturbed areas support multiplication of natural enemies and their migratation into crops fot biological control of pest population. Conservation of diversity in agroecosystem based on ecological studies of insect life can reduce pest problems. Maintenance of diverse cropping at different growth stages and utilization of diversed multiple management tactics will results in suppression of pests under broad-ranged stresses. These types of broad-ranged stresses create difficulties for the pests to locate their favourable crop hosts as well as to develop resistance rapidly against adopted pest management measures (Schillhorn et al. 1997; Pattison 2005).

Insect ecological studies also help to select various alternate host plants which can serve as trap and cover crops. Such crops, when intercropped or border-cultivated, not only recruit entomophagous insects in their battle against insect pests on major crops but also create a nice habitat for feeding and overwintering of beneficial insects. Between his raspberry rows, dandelions flower serve as source of food for nectar- seeking and polliferous insects. Insect ecological studies also laid the foundation of insect chemical ecology that yielded the discovery of so many semiochemicals and their potential implementation in pest management program of so many insect pests (Pattison 2005). For example, discovery of pheromones (methyl eugenol for fruit flies and gossyplure for pink bollworm), allomones, kairomones and synomones are based on insect chemical ecology studies (Saha and Dhaliwal 2012).

2.6 MANAGEMENT OF INSECT ECOSYSTEM: STRUCTURE TO IMPLEMENTATION

Ecosystem management influences the interaction of pests with other functional and structural components of ecosystem and determines the sustainability of pest management approaches. It is, therefore, imparative to understand the structure/components of the prevailing ecosystem (Schowalter 2011). Physical structure of an ecosystem represents the size and distribution of its densitydependent (biotic) and density-independent (abiotic) components. These both types of components determine the types and intensity of application of pest management tactics/strategy. For example, an ecosystem having biotic and abiotic factor very severe for pest survival should be exhibited with do-nothing or at least with minimum anthropogenic pest management tactics. Unlikely, if the ecosystem is characterized by such biotic and abiotic factors which favour the pest outbreak and population.

Another component of an ecosystem is the trophic structure which represent the numbers, mass (biomass), or energy content of organisms in each trophic level in form of numbers pyramids, biomass pyramids, or energy pyramids (Elton 1939). Trophic structure constitutes both herbivores and their predators as well parasites. The interaction of these herbivore and their natural enemies affect the complexity of herbivore and carnivore effects on ecosystem structure and function. If the trophic structures are favoring the herbivoric effects and limiting the carnivoric effects, then that trophic system support the implementation of pest management tactics. On the other hand, if the trophic structures are favoring the carnivoric effects and limiting the herbivoric effects, then that trophic system support the implementation of donothing strategy because predators, parasites and other natural fatal factors are actively contributing to regulate biological equilibrium (Schowalter 2011).

2.7 PEST MANAGEMENT AND ECONOMIC DECISION LEVELS: DECISION STAIRCASE, CONCEPTS AND PRACTICALITY

The decision staircase of pest management program shows that successful and sustainable pest management depend on certain pillars of pest management tactics that basically stand on the foundation of six slabs (biology, ecology, threshold, models, sampling and taxonomy) and one of those is economic decision levels (thresholds). Economic decision levels (EDLs) are indispensible for devising and implementing insect pest management program in an effective and economical way (Pedigo and Rice 2009). The comprehensive and true practical knowledge of such decision levels ensure the sensible and timely use of insecticides because these levels highlight the exact density of insect population that may cause economic damage if insecticides are not used. Ignorance of these economic decision levels leads to ridiculous economic gaffes spending more cost on pest management and crop protection than benefits a pest management strategy/ tactics may ensure. A comprehensive and proper knowledge, understanding and use of these economic decision levels can enhance the profit ratio of the growers and ensure the conservation of the environment and biodiversity (Dhaliwal et al. 2006; Pedigo and Rice 2009). Briefly, proper and sensible utilization of EDLs has following pluspoints (Knipling 1979; Pedigo and Rice 2009):

- Sensible use of insecticides and avoidance from the indiscriminate use of insecticides

- Reduction in insecticides use

- Increase producer's profit ratio

- Conserve natural biodiversity

- Conserve the environment quality

- Solution of some functional, biological and environmental issues like foodsafty and food-security, development of ecological-backlash in form of 3Rs (Resistance,

Resurgence and Replacement) and negative impacts on environment, human and various non-target organisms.

These EDLs include EIL (Economic Injury Level), ETL (Economic Threshold Level), GT (Gain Threshold) and DB (Damage Boundary). Among these, ETL is the practical operational level which is recommended to and being practiced by the growers for making pest management decisions in many situations. ETL is mostly used for making decision about the strategic implementation of curative/therapeutic management tactics including mainly insecticides. The use of ETL in pest management programs depend on following four decision rules (Knipling 1979):

2.7.1 No-Threshold Rule

No-threshold rule is applied when: i) sampling and scouting is uneconomical; ii) timely and practical implementation of remedy measures for the problem is difficult; iii) timely remedy and treatment of the problem is impractical; iv) too low ETL of pest; v) threats of quality losses, outbreak and transmission of fatal disease, too quick growth potential of pest/disease are confirmed); and vi) general equilibrium position of the pest always remains intensively above EIL.

2.7.2. Nominal Threshold Rule

This rule is established on the bais of skills, expertise and experiences of the entomologists. It is the most widely and frequently used economic threshold rule in any pest management program. An expert and professional entomologist defines ETLs for the prevailing pests based on his longlife field experiences.

2.7.3. Simple Threshold Rule

According to this rule, field experiments are conducted under controlled conditions for various infestation levels of a pest to determine its ETL. The ETL is calculated based on various parameters like market values of the commodity, cost of pest management practice, damage/loss done by specific pest density or pest infestation level and yield as well as monetary loss reduced by pest management practice on per plant or unit area basis. The ETLs calculated in this way are used in pest management.

2.7.4. Comprehensive Threshold Rule

This rule implies the ecological/environmental threshold levels. The determination and calculation of such economic threshold involves all possible interactive impacts of different biotic and abiotic stresses on tritrophic cascade of pest, plant and natural enemies. Calculattion and implementation of this comprehensive threshold is possible only if on-farm GPS and GIS based information collection and delivery setup is available.

2.8. PEST MONITORING AND SCOUTING

2.8.1. Pest Monitoring

Monitoring phytophagous insect pest and their natural enemies is the fundamental tool in IPM for making management decision (Dhaliwal and Arora 2003). Monitoring highlights the fluctuation in distribution and abundance of insect pests, outbreak and life history of insect pests and influence of biotic and abiotic factors on pest population. Monitoring helps in detection of outbreak of indigenous and exotic insect pest species, understanding ecological, climatological and biological factors regulating the pest movements, determining emergence pattern and generation peaks of important insect pests and detecting the rate of development of insecticide resistance in important insect pests. All these monitoring oriented information help to develop predictive models which are used to forewarn the growers regarding insect pest's outbreak and to develop and use sampling schemes as well as initiate management strategies (Dhaliwal and Arora 2003). Monitoring also provides following important information:

- ⊙ Kind and category of pests prevailing the crop

- ⊙ Density of the pest

- ⊙ Whether the prevailing pest density demands control measures? Have the implemented pest management practices suppressed the pest population significantly?

Various sampling tools including absolute estimate, relative estimate and population indices are used for monitoring insect pests. Various techniques which can be used in monitoring programs of various insect include in-situ counts, knockdown (jarring, shaking, beating, heating or chemical knockdown), netting, trapping (light traps, pheromone traps, pit-fall trap, Malaise trap, sticky traps), extraction from soil by sieving, washing, floatation, berlese funnels and soil-sampler (Pedigo and Rice 2009; Saha and Dhaliwal 2012).

2.8.2 Pest Scouting

Pest scouting is the inspection of a field to determine crop condition, severity of pests and their losses and density of beneficial fauna in a specific crop using already well defined and established pest-scouting methods/techniques. The philosophy and principles of pest control stress on the management of a pest only when its incidence is causing or is expected to cause damage above the acceptable/tolerable limit. The philosophy of pest management also emphasizes on the implementation of a pest management strategy which ensures a significant reduction in pest density to a tolerable level and encourages the conservation of non-target fauna of an agroecosystem (Dhaliwal and Arora 2003). Sustainably successful, effective and economical pest management program is always based on some important informations including: 1) which type of pest is prevailing in the system (pest identification), 2) What is the pest density per unit area? (ETL), 3) whether the pest density/population is increasing or decreasing (pest dynamics), 4) when the pest is present (pest activity period) and 5) How much damage the crop can tolerate (Host plant resistance). These informations can be obtained by implementing regular pest scouting program in the crop during crop season. An efficacious and successful integrated pest management (IPM) program absolutely depends on accurate identification of insect-pest. A correct identification of any insect pest can be accomplished by conducting regular pest-scouting. This practice, if performed accurately, helps in early prediction of pest outbreak, selection of appropriate control measures from the available widest range of pest management options and bringing satisfactory socio-economical and ecological, benefits. A successful pest scouting program depends on accurate history and record-keeping of conditions and locations of farm, pest incidence, pest mapping and pesticides usage. These secords facilitate the farm-managers or growers in anticipating crop conditions, diagnosing epidemic or endemic pest problems and tracking each field of the farm. Pest scouting helps in identifying the pest(s) problem(s), determining the exact site of pest problem, deciding either control is needed or not, selecting true insecticides, knowing the types of available control measure, determining and selecting more appropriate, effective and economical control measure, evaluating the evidences of effectiveness of control measures used and assessing their risks and benefits. It also demonstrate pest population trends, crop conditions, insect growth stages, current weather conditions, to-date degree of damage to crop, crop growth stage, expected yield, other pest problems, activity and incidence of natural enemies, prevailing economic decision levels of key pest and success of already use pest management measures All these informations help in understanding population dynamics of insect pests and their potential impacts on yield and cost-benefit-ratio (CBR), determining the either the pest management measures should be adopted or not?, selection of the types of control measure that is needed to prevent economic losses. Pest scouting determines either pest population is above or below ETL or insecticides should be used or not .

Timing of pest scouting is also very crucial principle which ensures the successful and result-oriented pest scouting program and ultimately guarantees the success of pest management strategy. Time of pest scouting vary from crop to crop, insect pest species to species, plant growth-stage to growth-stage and locality to locality. Prompt identification of existing pests and appropriate selection as well as timely application of pest control strategies can minimize economic impacts of pests on any crop. Crop scouting calendars based on previous records and data illustrate the timing of scouting associated with crop and insect-pests found in any specific locality. Consistent and frequent monitoring and scouting is very imperative because insectpest dynamics keep on fluctuating quickly and frequently throughout cropping season. Optimum plant populations are very critical for obtaining good yield; that's why crops scouting should be initiated within 1-2 weeks of plant emergence. Weekly scouting is appropriate early in the growing season. However, when insect pest population is approaching a control threshold, fields may require scouting daily; while, bi-weekly scouting is normally sufficient later in the season. For those insect pests which appear later in the season (like armyworms, aphids etc.) and may approach control thresholds in a matter of days when field and weather conditions favour these later-season pests, scouting should be continued weekly.

The crucial factors that laid down the foundation of accurate and successful pest scouting program include number of sampling and sampling patterns. The foundation of correct and accurate scouting program as well as of the collected data is laid on the appropriate number of sampling locations in a field that ultimately depends on field size, crop, pest type and stage of development, level of infestation, timing, etc. (Omafra 2009). Generally, the number of the recommended sampling locations for scouting insect-pests based on field size are 5, 8 and 10 locations for up to 8 (20 acres), 8-12 (20-30 acres) and 12-16 (30-40 acres) hectares, respectively. However, the fields larger than 16 ha (40 acres) should be split into units of 16 ha (40 acres) or less for insect pest scouting. Scouting pattern is also very important for accurate pest scouting of any insect pest. Use a scouting pattern that address all plant-growthregulating factors including changes in variety/hybrid, soil type, past cropping history, fertilizer/manure application insect pest etc. Following information and criteria should be considered for determining the type of scouting pattern:

⦿ The pest-scouting pattern should cover whole of the field areas andlocation for observation should be different for each time the field is scouted. However, already scouted locations/fields should be rechecked to monitor the latest situation of pest development when hot spots are recognized in crop production system. For the insect pests, which are uniformly distributed across the field [like, corn borers (Insecta: Lepidoptera; Pyralidae), rice and sugarcane borers (Insecta: Lepidoptera), bollworms (Insecta: Lepidoptera), jassids (Insecta: Homoptera; Cicadellidae), whiteflies (Insecta: Homoptera; Aleyrodidae) etc.], select sampling locations randomly and evenly from whole the field and start pest-scouting leaving field border area of at least 20 m (66 ft) width. Principally, avoid pest-scouting in headlands or from some surrounding rows at field edges and always practice the scouting evenly throughout the field.

⦿ For insect pests expected to develop in headlands or outside rows, select sampling locations randomly and uniformly at the boundaries of field and perform pest-scouting preferable around the edges of the field.

⦿ For insect pests, which usually develop in a particular field areas, sampling locations should be selected and pest-scouting should be concentrated on those specific locations. However, other areas of affected fields should also be scouted.

Lack of pest monitoring and scouting in the crops results in incorrect pest identification, inappropriate selection of pesticide, its dosage, application method and rapid development resistant in insect pest against insecticides.

2.9 SELECTION OF CONTROL PRACTICES

Principles and criteria Selection of an appropriate control practice or highly compatible control measures is a key to success of any pest management program. Following principles and criteria should be kept in mind while selecting single or set of control practices;

2.9.1 Feasibility in Available Resources

The selected control practice(s) should be feasible under resources available at farms. Principally, the availability of managerial time, adequate work-force/labor, as well as of appropriate and functional equipments, required for undertaking any particular pest management practice(s), should be predetermined. The availability or accessibility of above-mentioned farm resources are perhaps the major constraints in implementing prticular pest management strategies (Schillhorn et al. 1997)

2.9.2 Flexibility in Cropping Program

Before selecting control practice(s), it should be determined whether the cropping system under discussion has sufficient flexibility to respond and tolerate pests under the influence of specific pest control practice(s) (Knipling 1979; Inayatullah 1995).

2.9.3 Economic Feasibility

The economic feasibility of the pest management practice(s) should be estimated and compared. After the establishment of economic thresholds for many pests in the field, there is no hard and fast economical pest control decision rule for the selection of one or many pest control strategies. As a rule, economic feasibility can be determined on some economic basis. First, the expected benefits of a given pest management practice(s) should be considered and if the expected costs exceed over time, the pest management practice(s) should not be adopted. If several pest management practices have estimated benefits/returns higher than their expected costs, then select only those which prove economically and operationally more efficient and sustainable in crop production system. If more than one falls into this category, select the one with the greatest expected return. Costs of a pest management practice(s) include the value of any special equipment and machinery, work-force, pest-management inputs and managerial time/services needed.

2.10 GOALS OF PEST MANAGEMENT

Pest management measures should address the principal goals IPM program. The main goals of any pest control program/strategy include prevention (hinder pest outbreak from getting epidemic form), suppression (reducing density and/or damage of pest to a tolerable level) and eradication (complete destruction of pest).

Prevention may be a goal of any pest management program/strategy when the incidence or abundance of any pest is foreseeable and predictable in advance. For example, for management of mosquito, house flies, household insect pests, locust swarm etc. prevention is the best principal goals for successful suppression and control of pest population.

Suppression is a common goal of any pest management program/strategy in many pest situations. The philosophy of this specific goal is to suppress pest density and minimize the pest damage to tolerable and acceptable level. This goal addresses the philosophy of the holistic concept of IPM program.

Eradication is a common goal of any pest management program/strategy for in indoor areas and for those insect pests which infest food commodities or act as carriers/vectors of animal diseases where loss of single food-grain/commodity or life is not bearable. In indoor and closed areas, the achievement of this goal by any pest management program is comparatively easier than outdoor and opened areas because enclosed environment is comparatively less complex and easier to manage than outdoor areas/environment. For example, the prime goal of pest management program will be eradication of insect pests inside enclosed areas, like apartments, teaching institutions (schools, universities, colleges etc., offices, industrial units` because the incidence of certain pests and their losses cannot be tolerated there and zero-tolerance is set as basic principle and goal of pest management program.

2.10.1 Ecological Backlashes: Causes and Management

An effective pest management tactic suddenly loose its effectiveness or becomes totally ineffective. The failure of any pest management tactic suddenly or with the passage of time may be due to selection of inappropriate and incompatible tactics and improper application technique. However, if appropriate and compatible tactics are used with proper technique even then pest management tactic may face failure which is due the counter responses exhibited by the insect species against the stress imposed by that pest management tactic. Such counter responses are called ecological backlash which consist of three major sources of this phenomenon. These three sources include "three Rs" i.e., resistance, resurgence and replacement. An understanding of the philosophies of these sources of ecological backlashes also helps to make some decision on the use of any pest management tactic/strategy. If the pest management measure is facing the problem of ecological backlash, then use of more than one control measures is appropriate for delaying ecological backlash. For selection of appropriate pest management tactics/ strategy, the need is to understand and have comprehensive knowledge of all the factors which influence the rate of development of ecological backlash in insects. These factors are categorized into operation and biological factors. The operational factors include the prolonged exposure of insect population to control tactic, selection pressure on every generation of insect pest, very high selection pressure of control tactic, absence of functional refuges, coverage on large geographical area, continuous application of insecticides with same mode of action and setting of very low population threshold. Biological factors include exhibition of no or little migratory behavior by inset population, monophagous nature of feeding of pest population, short generation time of insect species and very high natality of insect species. The techniques which can slowdown the rate of development of ecological backlash include: 1) reducing selection pressure and conserve susceptible gene pool in population by moderation technique; 2) saturating insect defense mechanisms by doses of pest management tactic that can overcome the ecological backlash; 3) reducing the selection pressure by adopting multiple attacking system/approach (Knipling 1979; Pedigo and Rice 2009).

2.11 DEVELOPMENT AND IMPLEMENTATION OF IPM STRATEGY

After the reception of any pest management containing both preventive and therapeutic practices, there is need to develop IPM strategy which depends on the principle of combining tactics. For the development of IPM strategy, first identify potential preventive and therapeutic measures, evaluate the tactics individually, formulate conceptual plan of potential system, conduct field trials with the system to determine costs, compatibility of the tactics, effectiveness of the system and deploy successful program that offer on-farm flexibility (Schillhorn et al. 1997; Pedigo and Rice 2009). Other than these, there are also so many other steps and procedures that should be adopted to develop an IPM program/strategy having high success and sustainability rate. The step-by-step procedures for developing an IPM program/strategy are given below (Norris et al. 2002; Pedigo and Rice 2009):

- Identify all insect pests, their life stages as well as their natural enemies in the system.

- Establish first crude and then refines as well as improved monitoring and pest scouting guidelines

- Establish injury levels and action threshold for each pest species in the system

- Establish a record keeping system for evaluating and improving any IPM program

- Develop a list of acceptable management strategies for each pest preferably the preventive strategies and then therapeutic strategies.

- Develop a specific criterion for the selection of pest management methods like;

- Least destructive to natural control and beneficial fauna, least hazardous to human health, least toxic to non-target organisms, exhibit sustainable reduction of pest population, easy to carryout in the system and most cost-effective in the short- and long-term situation.

- Develop guidelines for the selection of pesticide every time.

- Evaluate the sustainability of the IPM program

After the development of IPM program, next step is to determine its implementing protocol which will ensure the removal of all the barriers expected to destroy the philosophy of IPM program. Following are some suggestions which help in overcoming the barriers and smooth implementation of IPM program (SP-IPM 2008; Pedigo and Rice 2009).

- ◉ Always initiate the IPM program on small scale as well as on one place addressing predetermines short-term objectives

- ◉ Do not change everything at once; rather retain to the maximum degree all the procedures already in use.

- ◉ Hare the IPM program with all the management personnels involved in day-to-day IPM process as soon as possible so that they can understand and support the program.

- ◉ Keep all the personnels informed about what is being planned, what is happening now, the expected outcomes and what will happen next.

- ◉ Identify benchmark objectives of IPM program and then build a reward system for the recognition of the personnels who are adopting IPM program in well manner

- ◉ Publicize the IPM program through field staff, personnels, communication media and interview session, website development.

- ◉ Involve the community by developing an advisory committee composed of interested IPM personnels and organizations.

2.12 FACTORS CAUSING FAILURE OF PEST MANAGEMENT STRATEGIES

The sustainability of any pest management strategy depends on exploring the causes of its failure rather than rejecting it at once. Following are the main causes of the failure of pest management strategy:

- ◉ Incorrect identification of insect pest species: It results in wrong selection of control measures and failure of whole pest management strategy.

- ◉ Selection of inappropriate control measures: It poses negative impacts on non-target beneficial organism but does not ensure the suppression of the target insect species. and failure of whole pest management strategy.

- ◉ Selection of incompatible control measures: This cause reduces the efficacy of the selected control measures due to their antagonistic effects on the control potentials/properties of each other and ultimately the designed strategy face failure.

- ◉ Selection of inappropriate application technique: Sometimes the control measures are very strong and effective but their application techniques reduce the effectiveness of the control measures and failure of the pest management strategy.

- Improper timing of application of control measures: If very effective control measures are not applied at the correct time, it also leads toward the failure of the strategy. For example, application of chemical control during the unfavorable climatic conditions- when there is expectation of rain or windstorm, application at noon etc. - will either yield no control or least control of the target insect pest species.

- Application of same tactics excessively: This practice results in rapid development of resistance in insect pests against pest management measures and ultimately causes prompt reduction in the effectiveness of the pest management strategy. Development of resistance in insect pest species against control measures: Development of resistance against control measure due to its excessive and blind use is the critical and vital factor which causes the failure of any pest management strategy.

- Adverse climatic conditions: Sometimes climatic conditions, which are adverse for the performance of pest management strategy and favor the pest population growth, prove a big hurdle in the implementation pest management strategy, reduce its effectiveness and sometime result in its failure.

- Use of incorrect dosage of pest control measure: Pest control measure, specifically chemical and biological control do not perform effectively if they are not applied at their recommended and effective dosages even though they are implemented with strong and productive strategy. Use of incorrect (under dose or over dose) dosage either results in futility of pest management strategy (in case of under dose application) or rapid development of resistance (in case of over-dose application) in insect pests against control strategy (FAO 2014).

2.13 PUBLIC AWARENESS, LONG-TERM COMMITMENT, PLANNING AND IMPROVEMENT: PROCEDURES AND STRATEGIES

Success of any strategy and plan depends upon its adaptability to the public sector. A consistent and sustainable application of any pest management strategy is very important to get the desired results. The sustainability of any pest management program/strategy vitally relies on public awareness, long-term commitment of the involved personnels and planning as well as improvement in the existing pest management program. Most important step which ensures the sustainability of any pest management is awaring and demonstrating public with advantages and application strategies of introduced and developed pest management program so that they can implement it in strategically accurate and economically effective way. There are different procedures or method which can be used for the public awareness about the pest management program. These methods include public awareness campaigns through paper and electronic media, involvement of various NGOs and farmer's field school (FFS) working at the grass-root levels of the farming community, organizing training workshops, symposia, conferences etc., and field demonstration of the technology. Creating public/stakeholder awareness is not enough guarantee for the long-term sustainability of the transferred pest management technology. Every stakeholder should be committed to follow all the awareness-campaignhighlighted pre-requisites which ensure the sustainability in the effectiveness of pest management program/strategy. If the IPM personnels/stakeholders do not remain committed to follow the pre-requisite of sustainability of IPM program in the longrun, the IPM program will reduce its effectiveness and sustainability sooner or later. The IPM personnels/stakeholders of any specific area, locality or region must committedly consider, implement and monitor calendar, methods and tactics, strategies and emerging flaw/issues of implemented IPM program. Then another aspect of sustainable program includes planning to diagnose the cause of emerging issues and address their possible solutions for the improvement and long-term sustainability of already existing IPM program in the scenario of evolving pest issues and new technologies rather exploring and designing new pest management program.

2.14 GENERAL PRINCIPLES OF CHEMICAL AND BIOPESTICIDES

All pesticides, being poisonous substances, can impose detriment effects on all living things; that's why, they must be used in a judicious way. The selected techniques for insecticide application should ideally be target oriented so that the non-target organisms as well as the environment can be protected from the lethal residual and deteriorating impacts of insecticides . Additionally, comprehensive information of the equipment used for insecticide application is also very indispensible for: 1) the development of an anticipated dexterity of operations; 2) the selection and approximation of the number and kind of equipments for the optimized used of the selected equipment and accurate coverage of the crop in least spell of time.

For the successful implementation of insecticide based pest control program, comprehensive knowledge of application technique, target, application time, coverage requirement, droplet size, calibration requirement, precautionary measures for handling, agitation and mixing requirement and equipment and tools used are required. The determination of all knowledge based on knowledge of pest problem, insecticides, formulations, technique and equipment. The knowledge of pest problem tells:

1. Location of the pest-insect that helps in defining the target;
2. The most susceptible stage that aids to decide the time of insecticide application; and mobility/dispersal behavior of the pest that helps to define coverage and droplet-size requirement.

The knowledge of the insecticides states:

1. Their mode of action that define the application technique required;

2. Their degree of phytotoxicity that define their calibration requirement; and their mammalian toxicity that determines the precautionary requirements for handling insecticides.

The knowledge of insecticide formulations tells:

1. The type of solubility that defines the agitation requirements of the insecticides; and

2. The methodology of their mixing with water or other solvent for tank-mixing that determine the suitable measures and tools required.

The knowledge of techniques and equipments states:

1. The procedures/methods and protocols for their operation and maintenance so that they can be operated in the field without any field difficulties;

2. Their capabilities that help to estimate the number of equipment needed; and the type of technique that should be selected and this information help to select the suitable and most appropriate equipments. The various factors or principles which determine the success or failure of pest control by insecticides are discussed below:

2.14.1 Selection of Suitable Application Technique

There are various techniques which are used for application of pesticides. The formulations of insecticides are made available in liquid, dust-powder, granule or slow-releasing forms which ensure their application in small quantities over large area. Application of insecticides in small recommended quantity is possible only if proper application technique is adopted. Therefore, selection and adoption of the most appropriate technique and equipment for pesticide application are very vital for depositing insecticides uniformly, performing pest control operation accurately and getting the effective results from any pesticides. The selection of pesticide application technique depends on the type, life-stage and feeding as well migratory behavior of the insect pest, site/substrate to be treated (on foliage, under the leaves, at root zone, plant whorl, breeding places etc), types of insecticide formulation, etc (Metthews 1979). For examples, granular insecticides against borers are applied by whorl application or by chemigation techniques but their foliar application will not yield the effective results. The most appropriate technique for the indoor control of mosquito is the use of chemicals in form slow-releasing or fogging technique. Effective monitoring and control of fruit flies is possible by pheromones when these are applied by trapping and insecticide coadministration technique. Similarly, for the control of eggs and young ones of mosquito and cockroach, the most appropriate technique is the breeding-site-treatment method for insecticide application. For cutworm control the most appropriate technique for the application of insecticides will the food-bait or chemigation technique. In short, it is the principle, while using insecticide, that the most appropriate technique should be used otherwise, the required results will not be achieved.

2.14.2 Selection of Appropriate and Quality Insecticides

Various types of insecticides like OPs, OCs, pyrethroids, carbamates, IGRs, neonecotenoids, biorationals and many others with novel mode of actions are used for the management of different insect pests, obtaining quality yield and reducing yield losses. But selection of an appropriate insecticide of standard quality (i.e., proper quantity of fresh not expired active ingredient with standard and good quality inert material in proper proportion as described on label) is first principle that will assure the success of insecticide based pest control program (Matthews 2000; Food and Agriculture Organization 2014)

2.14.3. Decision on the Timing of Application

After selection of an appropriate insecticide of standard quality, decision on the appropriate application timing is the next principle that can lay down the base of successful pest control program. Application of insecticide is very efficacious and yields the required result if applied at the most susceptible stage of the pest. The timing of insecticide application should prudently be considered and followed for good and economical results (Matthews 2000; Pedigo and Rice 2009).

2.14.4. Proper Application of Insecticides

Application of good quality insecticide at ideal timing does not yield good and economical results until or unless it is applied properly and qualitatively. The proper and quality application of insecticides is an imperative principle of pest control program. The proper and quality application of insecticides can be achieved if proper dosage is applied evenly, toxicant reaches the target, proper droplet size of insecticide is produces and sprayed and proper density of droplets is deposited on the target.

2.14.5 Application of Proper and Recommended Dosage

Every insecticide has its recommended dosage which is considered lethal for targeted insect pests but safe and non-phytotoxic for plants. The recommended dosage of selected insecticides should carefully be considered and applied to get good results. Over-dosage and under dose should be avoided because such practices do not yield the required results rather impose many destructive, hazardous and undesired impacts like, ecological backlash, environmental deterioration, phytotoxicity, health hazardous, deterioration of biodiversity etc.

2.14.6 Selection of the Most Appropriate Equipment and Associated Tools/ Materials

Insecticides are applied by different methods including foliar-application/spraying, dusting, drenching, fogging etc. Depending upon the application method recommend for the selected insecticides, the most appropriate equipment and associated tools especially nozzles should be selected for getting good and economical results. Improper selection of equipments and nozzles results in poor and uneconomical control of insect pests. Different equipments which are used for insecticides application include hydraulic, centrifugal and gaseous energy sprayers, aerosols sprayers, dispensers, foggers, dusting equipments, granular applicators etc. Similarly, various types of nozzles like hydraulic energy nozzles (hollow-cone, fan/flat-fan and impact/deflector/floodjet, adjustable/triple-action nozzles), thermal energy nozzles, gaseous energy nozzles, centrifugal nozzles etc. are used for pesticides application. For tank-mixing before spray different solvents like water, oils etc. are used. The proper selection of spraying equipment and type of nozzle is very vital for achieving good coverage and accurate droplet size and ultimately successful and economical control program.

2.14.7 Compatible and Synergistic

Selection of insecticides which don't have any antagonistic; rather have synergistic interaction with other insecticides ensure the success of multiple-attack-system. Therefore, those insecticides should be selected for multiple-attack-system that are compatible with each other and synergize the toxic effects of each other. The selection and application of insecticides having antagonistic effects on each other result in the failure of multiple-attack-system. The selected insecticides should also be compatible with control measures other than insecticides or they should be manipulated in such way as reducing the side effects on other control measures.

2.15 SUSTAINABLE PEST MANAGEMENT

A successful and sustainable pest management depends on the knowledge of the strategy, pest biology and pest ecology in agroecosystem. The development of an effective pest control strategy is reliant and contingent on ten pillars including:

- ⦿ Correct identification of the pest and its pest-status;
- ⦿ Determination of pest etl;
- ⦿ Knowledge of the available control tactics;
- ⦿ Selection of control tactic(s);
- ⦿ Decision of appropriate timing (when?), Conducive technique (how?) And targeting site (where?);
- ⦿ Determination and choice of the pest control goals;
- ⦿ Selection of effective pest monitoring tool/techniques;
- ⦿ Identification of factors causing failure of pest control tactics;
- ⦿ Public awareness and long-term commitment;
- ⦿ Planning for and improvement in in pest management strategies.

A comprehensive knowledge of various pest and ecosystem associated aspects like pest population ecology and dynamics, pest population structure and interactions and structure, function and regulators of ecosystem is compulsory for determining an appropriate pest management strategy. Based on aforementioned information, appropriate strategy can be selected and implemented from already well defined four strategies including:

1. "Do-Nothing Strategy",
2. "Reduce-Number Strategy",
3. "Reduce-Crop Susceptibility Strategy" and
4. "Integrated-Strategy" to cope with the emerging or prevailing pest condition.

These basic concepts and principles of sustainable pest management that are based on well-defined goals, estimates of population size, evaluation and comparison of available management options and monitoring as well as evaluation of practiced management activities/strategies regarding benefits and costs. Defined goals determine better and appropriate available management options in a prevailing situation; population estimates help in determining action threshold and deciding timeframe for initiation necessary actions. Effectiveness of available control options in stipulated time-period, environmental and social consequences and cost-benefit-ratio can help in ranking out the efficient, economical and ecofriendly management strategies. This chapter focuses on the fundamental concepts and principles, practical strategies and various techniques of insect pest management.

Pest management is an integral and vital component of manipulating, managing and regulating natural resources and agricultural systems. An area-wide public awareness campaign about the emerging pest issue must be organized and comprehensive knowledge of pests must be outreached to transform the aptitude, enhance the capacity and motivate the willingness of individuals to manage pests. An effective pest management entails a long-term and enduring commitment to pest management or pest eradication program by the industry groups, government entities, society and community. Discussion, meetings, consultation, entrepreneurship and partnership arrangements between industry groups, civic and rural communities, local governments and state government agencies must be established to attain a concerted, corporative and collaborative approach and strategy to pest management. Pest management or pest eradication planning, as per status of the emerging pest, must be reliable, consistent and sustainable at local, regional, state and national levels and must guarantee resources target urgencies, priorities and primacies for pest management recognized at each level. Preventative and anticipatory pest management is accomplished by early detection of the pest outbreak, local, regional, state and national level inhibition of pest migration, dispersal and spread as well as by intervention measures and strategies to control pests. Pest management must also be based on integration of highly compatible as well as ecologically and socially responsible therapeutic pest management practices and strategies that ensure environmental protection, food security, productive capacity of natural resources and conservation of natural resources as well as diversity.

Fundamental and applied research about emerging or major/key pests and consistence, systematic and regular monitoring and evaluation of pest control activities is essential and indispensable to improve pest management practices for their better sustainability in any pest management program and system.

Fig. : General Principles which lay down the foundation of successful and sustainable pest management.

3

INSECT PHYSIOLOGY

3.1 INTRODUCTION

Insects are the most diverse of all organisms on earth. Their general body plan allows for this tremendous diversification in form. Insects are arthropods meaning they have an external skeleton that covers the internal tissues. The exoskeleton protects the internal tissue but also allows for sensory systems to function. Insect physiology is the specialized study of how insects live and reproduce. A discussion follows of how the organ systems function in insects. The first section will be concerned with a description of the exoskeleton and the molting process involved in growth and development followed by sections describing the major organ systems of insects.

Insect physiology is the study of how insects live and reproduce. This is a historic area of research that continues today. The study of insect physiology is usually divided into a systems approach. These systems are the same required by all animals. The major systems are: digestive, excretory, circulatory, immune, muscular, nervous, and reproductive. All of these will be discussed to some extent in this module. Additional information can be obtained from other sources for readers requiring a more in depth examination of each system.

3.2 INSECT GROWTH, DEVELOPMENT, AND REPRODUCTION

3.2.1 Molting

An insect's skeleton is on the outside of its body and is called an exoskeleton. It serves as a support for muscles and internal organs as well as a covering. As the insect's rigid exoskeleton cannot expand much, it must be shed and replaced with a larger one as the insect grows. This process is called molting. The life stage between each molt is called an instar.

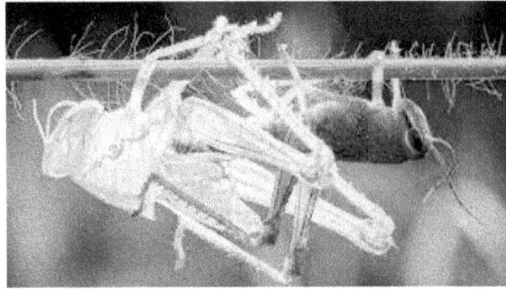

Molting is governed by hormones. Cuticle secretion and the molt cycle are controlled by ecdysone, a steroid hormone. The hormone is secreted by a gland in the thorax, which is in turn controlled by a hormone from the brain. Whenever the brain receives the appropriate stimulus, the insect will molt. A new cuticle forms under the old one, then the old exoskeleton splits and the insect wriggles its way out. Many insects eat their own discarded skin. The new cuticle is soft at first. The insect may swallow air to expand its own volume and stretch the new exoskeleton before in hardens, usually within about an hour.

Kinds of Insect Growth

Molting may occur up to three or four times or, in some insects, fifty times or more during its life. The two kinds of insect growth are directly related to molting patterns. Those that show a determinate pattern of growth have a fixed number of molts, whereas those that have indeterminate growth continue to molt indefinitely.

Classification of Insects According to Growth

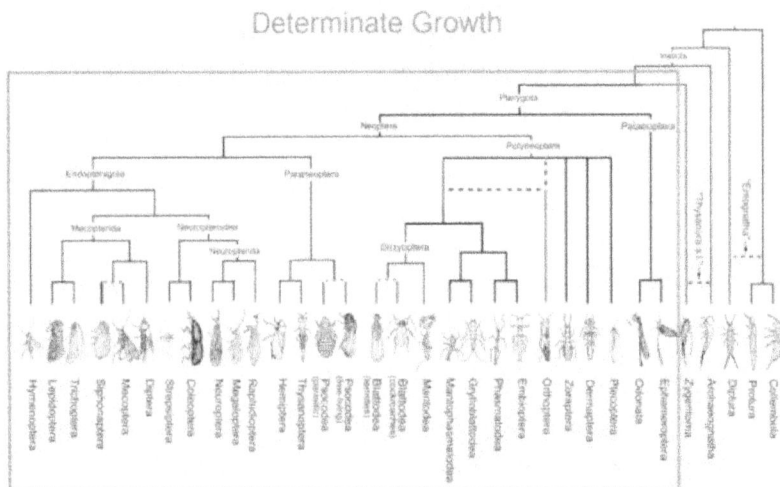

3.2.2 The Stages of Molting

1. Apolysis

Molting hormones are released into the haemolymph and the old cuticle separates from the underlying epidermal cells. The epidermis increases in size due tomitosis and then the new cuticle is produced. Enzymes secreted by the epidermal cells digest the old endocuticle, not affecting the old sclerotised exocuticle.

2. Ecdysis

This begins with the splitting of the old cuticle, usually starting in the midline of the thorax's dorsal side. The rupturing force is mostly from haemolymph pressure that has been forced into thorax by abdominal muscle contractions caused by the insect swallowing air or water. After this the insect wriggles out of the old cuticle.

3. Sclerotisation

After emergence the new cuticle is soft and this a particularly vulnerable time for the insect as its hard protective coating is missing. After an hour or two the exocuticle hardens and darkens. The wings expand by the force of haemolymph into the wing veins.

3.3 LIFE CYCLE

An insect's life-cycle can be divided into three types:

- ⊙ Ametabolous

- ⊙ No metamorphosis

- ⊙ These insects are primitively wingless where the only difference between adult and nymph is size. o Example: Order: Thysanura (Silverfish).

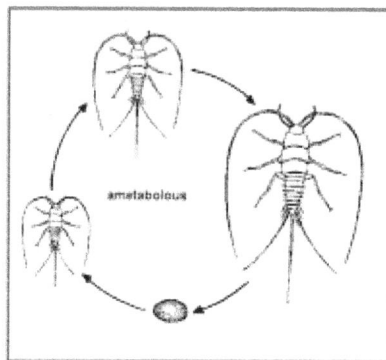

- ⊙ Hemimetabolous

- ⊙ Incomplete metamorphosis

- The terrestrial young are called nymphs and aquatic young are called naiads. Insect young are usually similar to the adult. Wings appear as buds on the nymphs or early instars. When the last moult is completed the wings expand to the full adult size.

- Example: Order: Odonata (Dragonflies).

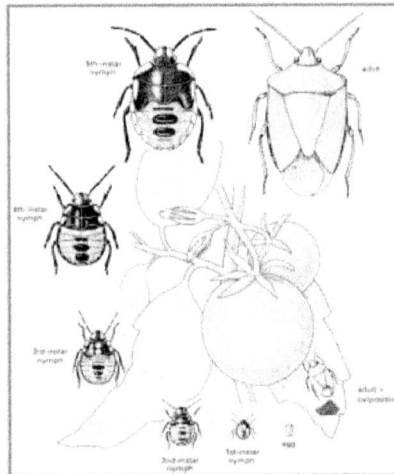

- Holometabolus

- Complete metamorphosis

- These insects have a different form in their immature and adult stages, have different behaviors and live in different habitats. The immature form is called larvae and remains similar in form but increases in size. They usually have chewing mouthparts even if the adult form mouth parts suck. At the last larval instar phase the insect forms into a pupa, it doesn't feed and is inactive, and here wing development is initiated, and the adult emerges.

- Example: Order: Lepidoptera (Butterflies and Moths).

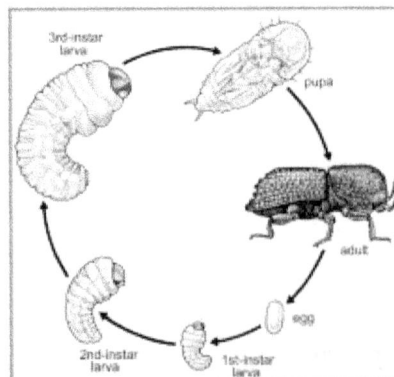

Classification of Insects According to Life Cycle

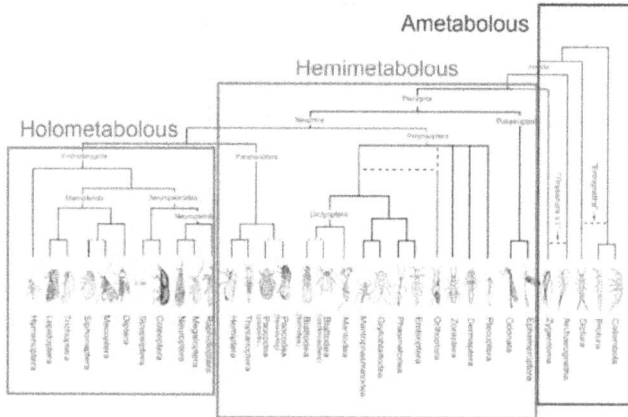

Insect development

- Instar = Stadium
- Imago = Adult

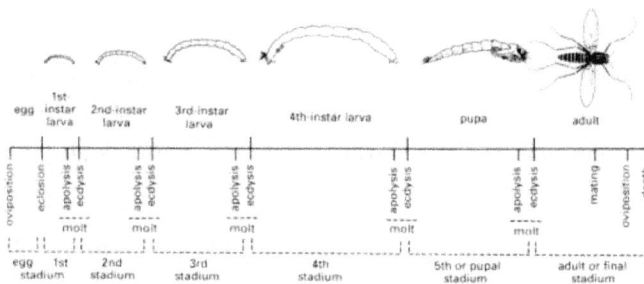

3.4 SEXUAL AND ASEXUAL REPRODUCTION

Most insects reproduce via sexual reproduction, i.e. the egg is produced by the female, fertilized by the male and oviposited by the female. Eggs are usually deposited in a precise microhabitat on or near the required food. However, some adult females can reproduce without male input. This is known as parthenogenesis and in the most common type of parthenogenesis the offspring are essentially identical to the mother. This is most often seen in aphids and scale insects.

3.5 METABOLIC SYSTEMS OF INSECTS

3.5.1 Insect External Anatomy

The insect is made up of three main body regions (tagmata), the head, thorax and abdomen. The head comprises six fused segments with compound eyes, ocelli, antennae and mouthparts, which differ according to the insect's particular diet, e.g. grinding, sucking, lapping and chewing. The thorax is made up of three segments: the pro, meso and meta thorax, each supporting a pair of legs which may also differ, depending on function, e.g. jumping, digging, swimming and running. Usually the middle and the last segment of the thorax have paired wings. The abdomen generally comprises eleven segments and contains the digestive and reproductive organs.

External Anatomy of a Grasshopper

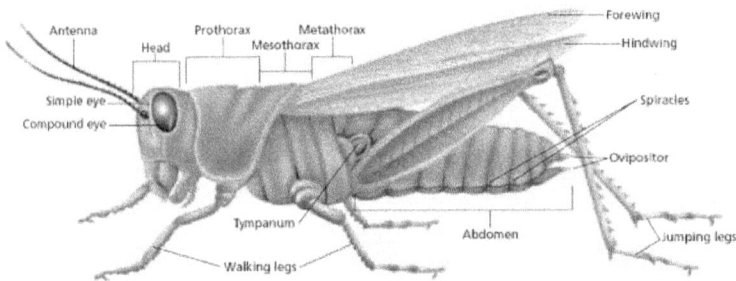

3.5.2 Muscular System

Many insects are able to lift twenty times their own body weight and may jump distances that are many times greater than their own length. This is not because they are strong but because they are so small. Muscle power is proportional to its cross-sectional area. Because the mass (the insect's body) that is moved is in proportion to its volume and the fact that they also have a better leverage system than humans do, they can jump remarkable distances.

The muscular system of insects ranges from a few hundred muscles to a few thousand. Unlike vertebrates that have both smooth and striated muscles, insects have only striated muscles. Muscle cells are amassed into muscle fibers and then into the functional unit, the muscle. Muscles are attached to the body wall, with attachment fibers running through the cuticle and to the epicuticle, where they can move different parts of the body including appendages such as wings. The muscle fiber has many cells with aplasma membrane and outer sheath or sarcolemma. The sarcolemma is invaginated and can make contact with the tracheole carrying oxygen to the muscle fiber. Arranged in sheets or cylindrically, contractile myofibrils run the length of the muscle fiber. Myofibrils comprising a fine actin filament enclosed between a thick pair of myosin filaments slide past each other instigated by nerve impulses.

Muscles can be divided into four categories:

Visceral: These muscles surround the tubes and ducts and produce peristalsis as demonstrated in the digestive system.

Segmental: Causing telescoping of muscle segments required for molting, increase in body pressure and locomotion in legless larvae.

Appendicular: Originating from either the sternum or the tergum and inserted on the coxae these muscles move appendages as one unit. These are arranged segmentally and usually in antagonistic pairs. Appendage parts of some insects, e.g. the galea and the lacinia of the maxillae, only have flexor muscles. Extension of these structures is by hemolymph pressure and cuticle elasticity.

Flight: Flight muscles are the most specialized category of muscle and are capable of rapid contractions. Nerve impulses are required to initiate muscle contractions and therefore flight. These muscles are also known as neurogenic or synchronous muscles. This is because there is a one to one correspondence between action potentials and muscle contractions. In insects with higher wing stroke frequencies the muscles contract more frequently than at the rate that the nerve impulse reaches them and are known as asynchronous muscles.

Flight has allowed the insect to disperse, escape from enemies, environmental harm, and colonize new habitats. One of the insect's key adaptations, the mechanics of flight differs from other flying animals because their wings are not modified appendages. Fully developed and functional wings occur only in adult insects. To fly, gravity and drag (air resistance to movement) has to be overcome. Most insects fly by beating their wings and to power their flight they have either direct flight muscles attached to the wings, or an indirect system where there is no muscle to wing connection and instead they are attached to a highly flexible box like thorax.

Direct flight muscles generate the upward stroke by the contraction of the muscles attached to the base of the wing inside the pivotal point. Outside the pivotal point the downward stroke is generated through contraction of muscles that extend from the sternum to the wing. Indirect flight muscles are attached to the tergum and sternum. Contraction makes the tergum and base of the wing pull down. In turn this movement lever the outer or main part of the wing in strokes upward. Contraction of the second set of muscles, which run from the back to the front of the thorax, powers the downbeat. This deforms the box and lifts the tergum.

3.5.3 Digestive System

An insect uses its digestive system to extract nutrients and other substances from the food it consumes. Most of this food is ingested in the form of macromolecules and other complex substances (such as proteins, polysaccharides, fats, and nucleic acids) which must be broken down by catabolic reactions into smaller molecules (i.e. amino acids, simple sugars, etc.) before being used by cells of the body for energy, growth, or reproduction. This break-down process is known as digestion.

The insect's digestive system is a closed system, with one long enclosed coiled tube called the alimentary canal which runs lengthwise through the body. The alimentary canal only allows food to enter the mouth, and then gets processed as it travels toward the anus. The insect's alimentary canal has specific sections for grinding and food storage, enzyme production and nutrient absorption. Sphincters control the food and fluid movement between three regions. The three regions include the foregut (stomatodeum), the midgut (mesenteron), and the hindgut (proctodeum).

In addition to the alimentary canal, insects also have paired salivary glands and salivary reservoirs. These structures usually reside in the thorax (adjacent to the foregut). The salivary glands produce saliva, the salivary ducts lead from the glands to the reservoirs and then forward through the head to an opening called the salivarium behind thehypopharynx; which movements of the mouthparts help mix saliva with food in the buccal cavity. Saliva mixes with food which travels through salivary tubes into the mouth, beginning the process of breaking it down.

The stomatedeum and proctodeum are invaginations of the epidermis and are lined with cuticle (intima). The mesenteron is not lined with cuticle but with rapidly dividing and therefore constantly replaced, epithelial cells. The cuticle sheds with every moult along with the exoskeleton. Food is moved down the gut by muscular contractions called peristalsis.

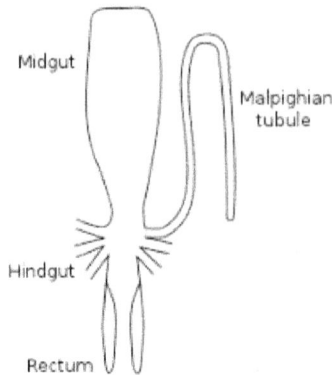

Stylised diagram of insect digestive tract showing Malpighian tubule (Orthopterantype)

1. Stomatodeum (foregut)

This region stores, grinds and transports food to the next region. Included in this are the buccal cavity, the pharynx, the oesophagus, the crop (stores food), and proventriculus or gizzard (grinds food). Salivary secretions from the labial glands dilute the ingested food. In mosquitoes (Diptera), which are blood-feeding insects, anticoagulants and blood thinners are also released here.

2. Mesenteron (midgut)

Digestive enzymes in this region are produced and secreted into the lumen and here nutrients are absorbed into the insect's body. Food is enveloped by this part of the gut as it arrives from the foregut by the peritrophic membrane which is a mucopolysaccharide layer secreted from the midgut's epithelial cells. It is thought that this membrane prevents food pathogens from contacting the epithelium and attacking the insects' body. It also acts as a filter allowing small molecules through, but preventing large molecules and particles of food from reaching the midgut cells. After the large substances are broken down into smaller ones, digestion and consequent nutrient absorption takes place at the surface of the epithelium. Microscopic projections from the mid-gut wall, called microvilli, increase surface area and allow for maximum absorption of nutrients.

3. Proctodeum (hindgut)

This is divided into three sections; the anterior is the ileum, the middle portion, the colon, and the wider, posterior section is the rectum. This extends from the pyloric valve which is located between the mid and the hindgut to the anus. Here absorption of water, salts and other beneficial substances take place before excretion. Like other animals, the removal of toxic metabolic waste requires water. However, for very small animals like insects, water conservation is a priority. Because of this, blind-ended ducts called Malpighian tubules come into play. These ducts emerge as evaginations at the anterior end of the hindgut and are the main organs of osmoregulation and excretion. These extract the waste products from the haemolymph, in which all the internal organs are bathed). These tubules continually produce the insect's uric acid, which is transported to the hindgut, where important salts and water are reabsorbed by both the hindgut and rectum. Excrement is then voided as insoluble and non-toxic uric acid granules. Excretion and osmoregulation in insects are not orchestrated by the Malpighian tubules alone, but require a joint function of the ileum and/or rectum.

3.5.4 Circulatory System

The main function of insect blood, hemolymph, is that of transport and it bathes the insect's body organs. Making up usually less than 25% of an insect's body weight, it transports hormones, nutrients and wastes and has a role in osmoregulation, temperature control, immunity, storage (water, carbohydrates and fats) and skeletal function. It also plays an essential part in the molting process. An additional role of the hemolymph in some orders, can be that of predatory defense. It can contain unpalatable and malodourous chemicals that will act as a deterrent to predators.

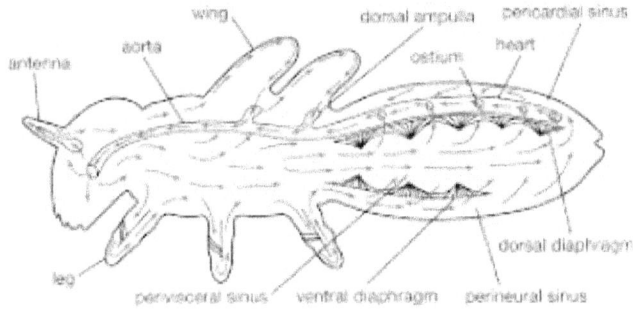

Hemolymph contains molecules, ions and cells. Regulating chemical exchanges between tissues, hemolymph is encased in the insect body cavity or haemocoel. It is transported around the body by combined heart (posterior) and aorta (anterior) pulsations which are located dorsally just under the surface of the body. It differs from vertebrate blood in that it doesn't contain any red blood cells and therefore is without high oxygen carrying capacity, and is more similar to lymph found in vertebrates.

Body fluids enter through one way valvedostia which are openings situated along the length of the combined aorta and heart organ. Pumping of the hemolymph occurs by waves of peristaltic contraction, originating at the body's posterior end, pumping forwards into the dorsal vessel, out via the aorta and then into the head where it flows out into the haemocoel. The hemolymph is circulated to the appendages unidirectionally with the aid of muscular pumps or accessory pulsatile organs which are usually found at the base of the antennae or wings and sometimes in the legs. Pumping rate accelerates due to periods of increased activity. Movement of hemolymph is particularly important for thermoregulation in orders such as Odonata, Lepidoptera, Hymenoptera and Diptera.

3.5.6 Respiratory System

Insect respiration is accomplished without lungs using a system of internal tubes and sacs through which gases either diffuse or are actively pumped, delivering oxygen directly to tissues that need oxygen and eliminate carbon dioxide via their cells. Since oxygen is delivered directly, the circulatory system is not used to carry oxygen, and is therefore greatly reduced; it has no closed vessels (i.e., no veins or arteries), consisting of little more than a single, perforated dorsal tube which pulses peristaltically, and in doing so helps circulate the hemolymph inside the body cavity.

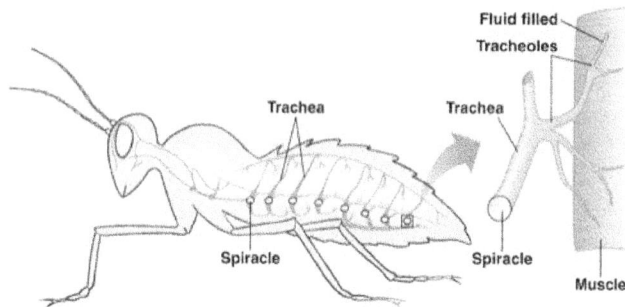

Air is taken in through spiracles, openings which are positioned laterally in the pleural wall, usually a pair on the anterior margin of the meso and meta thorax, and pairs on each of the eight or less abdominal segments, Numbers of spiracles vary from 1 to 10 pairs. The oxygen passes through the tracheae to the tracheoles, and enters the body by the process of diffusion. Carbon dioxide leaves the body by the same process.

The major tracheae are thickened spirally like a flexible vacuum hose to prevent them from collapsing and often swell into air sacs. Larger insects can augment the flow of air through their tracheal system, with body movement and rhythmic flattening of the tracheal air sacs. Spiracles are closed and opened by means of valves and can remain partly or completely closed for extended periods in some insects, which minimizes water loss.

There are many different patterns of gas exchange demonstrated by different groups of insects. Gas exchange patterns in insects can range from continuous, diffusive ventilation, to discontinuous gas exchange.

Terrestrial and a large proportion of aquatic insects perform gaseous exchange as previously mentioned under an open system. Other smaller numbers of aquatic insects have a closed tracheal system, for example, Odonata, Tricoptera, Ephemeroptera, which have tracheal gills and no functional spiracles. Endoparasitic larvae are without spiracles and also operate under a closed system. Here the tracheae separate peripherally, covering the general body surface which results in a cutaneous form of gaseous exchange. This peripheral tracheal division may also lie within the tracheal gills where gaseous exchange may also take place.

3.5.7 Endocrine System

Hormones are the chemical substances that are transported in the insect's body fluids (hemolymph) that carry messages away from their point of synthesis to sites where physiological processes are influenced. These hormones are produced by glandular, neuroglandular and neuronal centers. Insects have several organs that produce hormones, controlling reproduction, metamorphosis and molting. It has been suggested that a brain hormone is responsible for caste determination in termites and diapauses interruption in some insects.

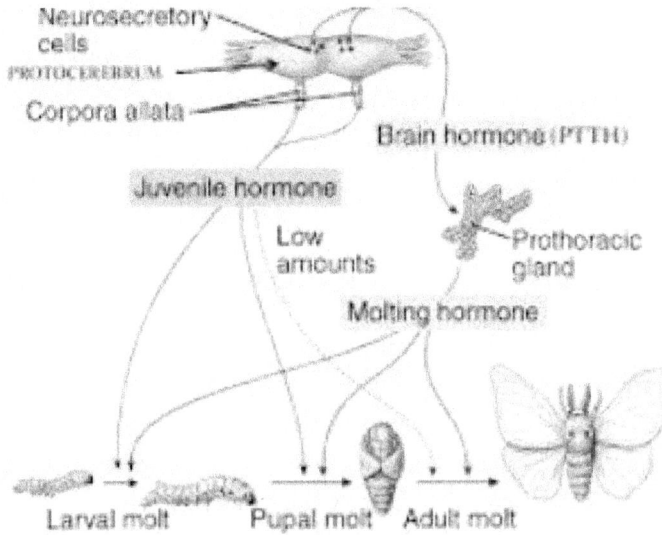

Four endocrine centers have been identified:

1. Neurosecretory cells in the brain can produce one or more hormones that affect growth, reproduction, homeostasis and metamorphosis.

2. Corpora cardiaca are a pair of neuroglandular bodies that are found behind the brain and on either sides of the aorta. These not only produce their own neurohormones but they store and release other neurohormones including PTTH prothoracicotropic hormone (brain hormone), which stimulates the secretory activity of the prothoracic glands, playing an integral role in molting.

3. Prothoracic glands are diffuse, paired glands located at the back of the head or in the thorax. These glands secrete an ecdysteroid called ecdysone, or the moulting hormone, which initiates the epidermal moulting process. Additionally it plays a role in accessory reproductive glands in the female, differentiation of ovarioles and in the process of egg production.

4. Corpora allata are small, paired glandular bodies originating from the epithelium located on either side of the foregut. They secrete the juvenile hormone, which regulate reproduction and metamorphosis.

3.5.8 Nervous System

Insects have a complex nervous system which incorporates a variety of internal physiological information as well as external sensory information. As in the case of vertebrates, the basic component is the neuron or nerve cell. This is made up of a dendrite with two projections that receive stimuli and an axon, which transmits information to another neuron or organ, like a muscle. As for vertebrates, chemicals (neurotransmitters such as acetylcholine and dopamine) are released at synapses.

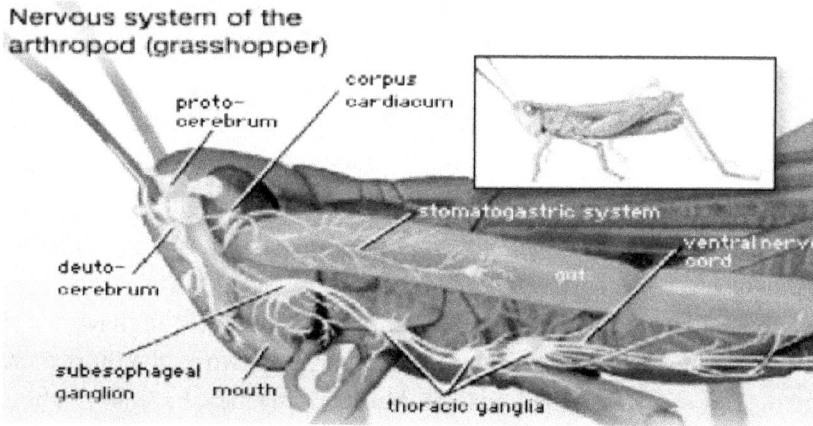

Nervous system of the arthropod (grasshopper)

3.5.9 Central Nervous System

An insect's sensory, motor and physiological processes are controlled by the central nervous system along with the endocrine system. Being the principal division of the nervous system, it consists of a brain, a ventral nerve cord and a subesophageal ganglion. This is connected to the brain by two nerves, extending around each side of the oesophagus.

The brain has three lobes:

- Procerebrum, innervating the compound eyes and the ocelli

- Deutocerebrum, innervating the antennae

- Tritocerebrum, innervating the foregut and the labrum.

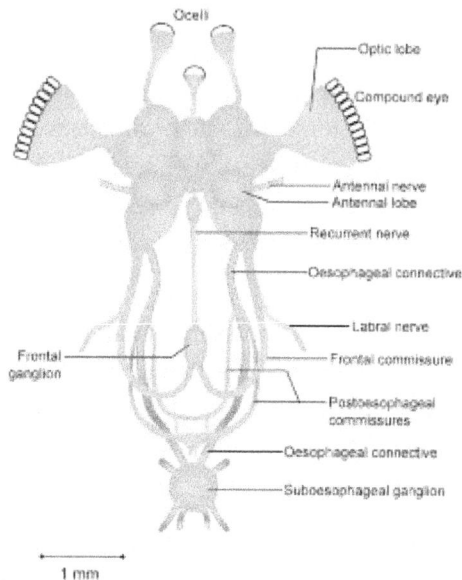

Anterior (frontal) view of the brain of the locust

The ventral nerve cord extends from the suboesophageal ganglion posteriorly. A layer of connective tissue called the neurolemma covers the brain, ganglia, major peripheral nerves and ventral nerve cords.

The head capsule (made up of six fused segments) has six pairs of ganglia. The first three pairs are fused into the brain, while the three following pairs are fused into the subesophageal ganglion. The thoracic segments have one ganglion on each side, which are connected into a pair, one pair per segment. This arrangement is also seen in the abdomen but only in the first eight segments. Many species of insects have reduced numbers of ganglia due to fusion or reduction.[8] Some cockroaches have just six ganglia in the abdomen, whereas the wasp Vespa crabro has only two in the thorax and three in the abdomen. And some, like the house fly Musca domestica, have all the body ganglia fused into a single large thoracic ganglion. The ganglia of the central nervous system act as the coordinating centers with their own specific autonomy where each may coordinate impulses in specified regions of the insect's body.

Peripheral nervous system

This consists of motor neuron axons that branch out to the muscles from the ganglia of the central nervous system, parts of the sympathetic nervous system and the sensory neurons of the cuticular sense organs that receive chemical, thermal, mechanical or visual stimuli from the insects environment. The sympathetic nervous system includes nerves and the ganglia that innervate the gut both posteriorly and anteriorly, some endocrine organs, the spiracles of the tracheal system and the reproductive organs.

Sensory Organs

Chemical senses include the use of chemoreceptors, related to taste and smell, affecting mating, habitat selection, feeding and parasite-host relationships. Taste is usually located on the mouthparts of the insect but in some insects, such as bees, wasps and ants, taste organs can also be found on the antennae. Taste organs can also be found on the tarsi of moths, butterflies and flies. Olfactory sensilla enable insects to smell and are usually found in the antennae. Chemoreceptor sensitivity related to smell in some substances is very high and some insects can detect particular odors that are at low concentrations miles from their original source.

Mechanical senses provide the insect with information that may direct orientation, general movement, flight from enemies, reproduction and feeding and are elicited from the sense organs that are sensitive to mechanical stimuli such as pressure, touch and vibration. Hairs (setae) on the cuticle are responsible for this as they are sensitive to vibration touch and sound.

Hearing structures or tympanal organs are located on different body parts such as, wings, abdomen, legs and antennae. These can respond to various frequencies ranging from 100 to 240 kHz depending on insect species. Many of the joints of the insect have tactile setae that register movement. Hair beds and groups of small hair like sensilla, determine proprioreception or information about the position of a limb, and are found on the cuticle at the joints of segments and legs. Pressure on the body wall or strain gauges are detected by the campiniform sensilla and internal stretch receptors sense muscle distension and digestive system stretching.

The compound eye and the ocelli supply insect vision. The compound eye consists of individual light receptive units called ommatidia. Some ants may have only one or two however dragonflies may have over 10,000, the more ommatidia the greater the visual acuity. These units have a clear lens system and light sensitive retina cells. By day, the image flying insects receive is made up of a mosaic of specks of differing light intensity from all the different ommatidia. At night or dusk, visual acuity is sacrificed for light sensitivity. The ocelli are unable to form focused images but are sensitive mainly, to differences in light intensity.

Color vision occurs in all orders of insects. Generally, insects see better at the blue end of the spectrum than at the red end. In some orders sensitivity ranges can include ultraviolet.

A number of insects have temperature and humidity sensors and insects being small and cool more quickly than larger animals. Insects are generally considered cold-blooded orectothermic, their body temperature rising and falling with the environment. However, flying insects raise their body temperature through the action of flight, above environmental temperatures.

The body temperature of butterflies and grasshoppers in flight may be 5°C or 10°C above environmental temperature, however moths and bumblebees, insulated by scales and hair, during flight, may raise flight muscle temperature 20–30°C above the environment temperature. Most flying insects have to maintain their flight muscles above a certain temperature to gain power enough to fly. Shivering, or vibrating the wing muscles allow larger insects to actively increase the temperature of their flight muscles, enabling flight.

Until very recently, no one had ever documented the presence of nociceptors (the cells that detect and transmit sensations of pain) in insects, though recent findings of nociception in larval fruit flies challenges this and raises the possibility that some insects may be capable of feeling pain.

3.5.11 Reproductive System

Most insects have a high reproductive rate. With a short generation time, they evolve faster and can adjust to environmental changes more rapidly than other slower breeding animals. Although there are many forms of reproductive organs in insects, there remains a basic design and function for each reproductive part. These individual parts may vary in shape (gonads), position (accessory gland attachment), and number (testicular and ovarian glands), with different insect groups.

Female

The female insect's main reproductive function is to produce eggs, including the egg's protective coating, and to store the male spermatozoa until egg fertilisation is ready. The female reproductive organs include, paired ovaries which empty their eggs (oocytes) via the calyces into lateral oviducts, joining to form the common oviduct. The opening (gonopore) of the common oviduct is concealed in a cavity called the genital chamber and this serves as a copulatory pouch (bursa copulatrix) when mating. The external opening to this is the vulva.

Often in insects the vulva is narrow and the genital chamber becomes pouch or tube like and is called the vagina. Related to the vagina is a saclike structure, the spermatheca, where spermatozoa are stored ready for egg fertilization. A secretory gland nourishes the contained spermatozoa in the vagina.

Egg development is mostly completed by the insect's adult stage and is controlled by hormones that control the initial stages of oogenesis and yolk deposition. Most insects are oviparous, where the young hatch after the eggs have been laid.

Insect sexual reproduction starts with sperm entry that stimulates oogenesis, meiosis occurs and the egg moves down the genital tract. Accessory glands of the female secrete an adhesive substance to attach eggs to an object and they also supply material that provides the eggs with a protective coating. Oviposition takes place via the female ovipositor.

Male

The male's main reproductive function is to produce and store spermatozoa and provide transport to the reproductive tract of the female. Sperm development is usually completed by the time the insect reaches adulthood. The male has two testes, which contain follicles in which the spermatozoa are produced. These open separately into the sperm duct or vas deferens and store the sperm. The vas deferentia then unite posteriorly to form a central ejaculatory duct this opens to the outside on an aedeagusor a penis. Accessory glands secrete fluids that comprise the spermatophore. This becomes a package that surrounds and carries the spermatozoa, forming a sperm-containing capsule.

The production of behaviorally appropriate movements requires coordinating neural interactions within an animal's nervous system, the integration of sensory input from the periphery, and instruction from central neuromodulatory commands.

3.6 INSECT COORDINATION AND INTEGRATION

Insect Flight

Insect flight is the most energy-demanding activity of animals. It requires the coordination and cooperation of many tissues, with the nervous system and neurohormones controlling the performance and energy metabolism of muscles, and of the fat body, ensuring that the muscles and nerves are supplied with essential fuels throughout flight. Muscle metabolism can be based on several different fuels, the proportions of which vary according to the insect species and the stage in flight activity. Octopamine, which acts as neurotransmitter, neuromodulator or neurohormone in insects, has a central role in flight. It is present in brain, ventral ganglia and nerves, supplying peripheral tissues such as the flight muscles, and its concentration in hemolymph increases during flight. Octopamine has multiple effects during flight in coordinating and stimulating muscle contraction and also energy metabolism partly by activating phosphofructokinase via the glycolytic activator, fructose 2,6-bisphosphate. One important muscle fuel is trehalose, synthesized by the fat body from a variety of precursors, a process that is regulated by neuropeptide hormones. Other fuels for flight include proline, glycerol and ketone bodies.

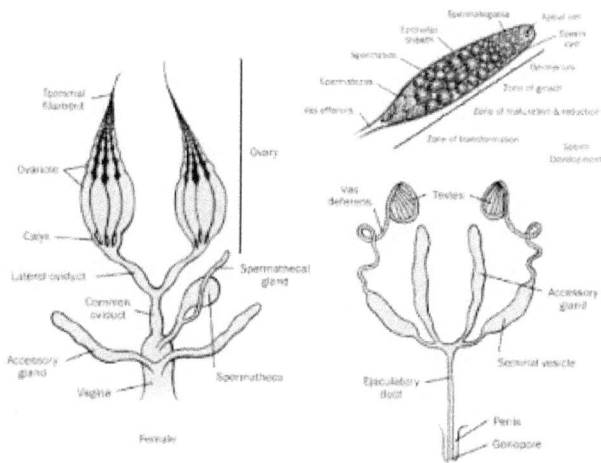

Insect Path Integration

Well before the advent of GPS, humans have had a need to revisit specific sites, for instance, the home cave or a specific tree that is abundant in fruit. While a map would be useful in these instances to avoid getting lost, creating a map requires information of where one is relative to the goal or the starting point of a trip. This is particularly difficult if there are no conspicuous features in the environment, for example in the desert or at sea. Early sailors, when venturing out into the sea, found a solution to this problem. They regularly updated their position relative to the point of departure by measuring traveling speed and direction, a strategy called dead reckoning. In animals this behavior is termed path integration. Many arthropods, including a wide variety of insects — for example, shield bugs, field crickets, cockroaches, flies, honeybees and ants — are known to use this strategy to return to their nest, hive or burrow by the shortest possible route after convoluted foraging trips. This is critical for survival since it reduces predatory pressure and allows the animal to minimize exposure to hostile weather conditions. In principle, during path integration animals continuously keep track of the distance and the directions traveled and then integrate this information to produce a single 'home' vector that takes them directly back to the point of origin.

Fig. :

This figure is a classic example of path integration from the foraging journey of an individual Saharan desert ant, Cataglyphis fortis. An ant departs the nest and finds food following a tortuous journey after travelling 354.50 m (blue path). Upon finding food the ant returns home directly (orange path) travelling 113.20 m.

4

BENEFICIAL INSECTS AND THEIR VALUE TO AGRICULTURE

4.1 INTRODUCTION: AN OVERVIEW

Beneficial insects provide regulating ecosystem services to agriculture such as Pollination and the natural regulation of plant pests. It aims to enhance insect-derived ecosystem services from a conservation perspective (i.e. enhancing beneficial insects in agricultural landscapes that provide ecosystem services to crops. Human cultures and civilizations have been maintained in countless ways through these beneficial insects, they regulate the pest population of many harmful pest species, produce natural products, and they also dispose the waste and recycle the organic nutrients. It should be considered in Thought that how much we depend on them for our survival and what kind of life would be without insects. Beneficial insects provide natural ecosystem services such as biological control of pests, soil formation, nutrient cycling and pollination of plants. Beneficial insects include pollinators important in the essential pollination process of all plants, and natural enemies of pests such as parasitoids and predators which are important in the suppression of pest damage to crops. Knowledge on management techniques to attract beneficial insects in the agricultural fields is a way forward to enhance agro ecosystems for increased crop production. Therefore, proper understanding and identification of natural

There are many insects found on agriculture land those are not threat to the crop production but beneficial to the farmers in different aspects, as Natural enemies, Pollinators, productive insects, Scavengers, weed killer and Soil builders. In present scenario the motive of the farmers is single sided, to gain only maximum profit, ignoring the impact on the beneficial insects, environment and human health. Insecticide can be a important crop production tool to maximize yield but Heavy and indiscriminate use of chemicals also exposes farmers to serious health risks, resulted in negative consequences for the insect those are beneficial to the farmers.

enemies, as well as pollinators in agricultural fields, is essential in promoting biological control and pollination activity. Natural enemies and pollinators, within legume fields, play a key role in ensuring sustainable production, especially in smallholder farms. There is a limited understanding of beneficial insects and the ecosystem services they offer to the agricultural production process.

4.2 REQUIREMENTS FOR ENHANCING BENEFICIAL INSECTS

The generalized intensification of agriculture and the use of broad-spectrum pesticides decrease the diversity of natural enemy populations and increase the likelihood of pest out breaks. Indeed, pesticide use has been shown to be associated with a large decrease in natural pest control services. Thus, enhancement of agro ecosystem appears to be one of the best ways in which we can decrease the use of chemical pesticides for pest and disease control. And it will increase the sustainability of crop production.

4.3 ROLE OF BENEFICIAL INSECTS AS

4.3.1 Pollinators

Insect pollinators are flower visiting Insects that forage on flowering plants to obtain plant-provided food (nectar, pollen). Flower-visiting insects have the potential to transfer male gametes (contained in pollen) to the female gametes while foraging, resulting in pollination. Insect-mediated pollination is an essential step in reproduction for the majority of the world's flowering plants, including numerous cultivated plant species i.e. Sunflower, Cucurbitaceous vegetables, Alfalfa, Coriander, Cardmom, Gingelly, Apple etc. Many crops depend on pollination for seed production and fruit set to achieve good yield. Globally, an estimated 35% of crop production is a result of insect pollination.

The Apis meliffera L. (European honey bee) is responsible for the pollination servicesin majority of crops. Non-Apis bees also are important pollinators of crops, especially for crops in which honey bees are inefficient pollinators (e.g. alfalfa, squash). A few non-Apis species are managed for crop pollination. Examples of managed non-Apis species include bumble bees, Bombus impatiens Cresson (Hymenoptera: Apidae) managed for cranberry (Vaccinium spp.) and greenhouse tomato (Solanum lycopersicum L.) pollination. Although bees are considered the most effective insect-pollinator of most plant species, other insects have been recognized for their contributions to pollination.

Flower visiting flies (Diptera) have been documented as proficient pollinators of several crops including carrot (Dacus carota L.), mustard (Brassica spp.), leek, (Allium ampeloprasum L.), and almond (Prunis dulcis). Weevil Elaeidobius kamerunicus (Coleoptera: Curculionidae) plays great role in pollination of Oil palm. Fig wasps are responsible for the pollination in both Smyra and Capri Fig Plantation.

4.3.2 Natural Enemies

Insect predators and parasitoids that attack and feed on other insects, particularly on insect pests of plants are considered natural enemies. Through this type of feeding, natural enemies contribute to a type of pest regulation referred to as natural biological control. Natural enemies' responsible Predaceous natural enemies belong to several insect orders and are generally characterized as free-living, mobile, larger than their insect prey, and are able consume several preys throughout their life cycle. But the parasitoids mainly belong to two orders Hymenoptera and Diptera, and their host ranges are considered to be more specialized than that of predator. Free-living adult parasitoids seek out a host, and depending on the parasitoid species, parasitize different life stages of their host (i.e., egg, larva, and pupa, adult). Parasitoids can lay an egg (solitary) or several eggs (gregarious) on or within their host and the immature parasitoid(s) feed on their host to complete development, kill their host, and emerge as free-living adult. In agricultural landscapes, natural enemies have the potential to prevent crop pests from reaching economically damaging levels (table-1).

Predators and parasitoids can suppress or delay pest population growth by contributing to pest mortality that is most vulnerable to herbivores. When diverse populations of natural enemies are present, pest control became more effective due to differing phenology.

Beyond natural biological control, natural enemies can be manipulated as part of integrated pest management programs through the importation and establishment of exotic natural enemy species (classical biological control), direct manipulation of populations (augmentative biological control), and, more pertinent to this research, through manipulation of their environment (conservation biological control)

4.3.3. Weed Killers

So many insects feed upon unwanted weeds just the same manner they do with the cultivated crops. In many cases the occurrence of these insects has contributed much towards eradication of the weeds.

4.3.4 Soil Builders

Insects which live in soil make tunnels, creating channels for smaller organisms, water, air, and roots to travel through. Insects improves soil aeration, and earthworm activity can enhance soil nutrient cycle, the soil physical properties, such as soil structure and tilth and activity of other beneficial soil organisms. Small Dung beetles makes tunnel walls with dung and also make dung balls that helps in maintaining the quality of the soil. Excreta of insets also enrich the soil. Examples- Beetles, Ants, Cut-worms, Larvae of flies, Crickets, Termites, Wasps etc.

4.3.5 Scavengers

Insects which feed on dead and decaying matter of plants and animals are called as scavengers. Insects (scavengers and decomposers) help in the biochemical cycling of the nutrients. Examples: Bark beetle, water scavenger beetle, Termites, Ants etc.

4.4 SOME BENEFICIAL INSECTS

- ⦿ Honey bee,
- ⦿ Silk worm larvae,
- ⦿ Lac insect,
- ⦿ Assassin bug,
- ⦿ Hover fly,
- ⦿ Aphidius calamani,
- ⦿ Syrphid fly,
- ⦿ Zygogramma bicolorata,
- ⦿ Termites,
- ⦿ Dragon fly,
- ⦿ Praying mantis,
- ⦿ Trichogramma sp.,
- ⦿ Damselfly,
- ⦿ Coccinella sp.,
- ⦿ Chrysoperla carnea.

4.5 SOME PRODUCTS FROM BENEFICIAL INSECTS

From thousands of years *Apis meliffera* L. (Honey bees) are important for gaining Honey and bee wax. And honey was the only sweetener, viscous fluid, produced by honeybees. It is collected from nectar from nectories at base flowers. Also collected from nectar secreted by plant parts other than flowers known as extra floral nectories. It is also collected from fruit juice, cane juice etc. In present, the developing markets are available for the other two products (Bee pollen and royal jelly) from honey. The bee pollen collects by pollen trap from ingoing pollen foragers. It is rich protein source. Bee pollen is a "complete" and good supplement in diet. It is available in health food stores. The royal jelly is secreted by gland of nurse bees when the glands are fully active. It is very nutritious food and is fed to the young workers larvae and queen larvae and adult. Royal jelly is milky and light pale in color. And it is also a good ingredient of some expensive skin care products, which helps in reducing wrinkles and works as anti-aging.

4.5.1 Production of Silk

A unique natural fiber silk cloth, which usually derives from silkworm, Bombyx mori. This "domestic" silk is famous for its finishing and light colors. The silk can also harvest from the many other species e. i. Antherea spp., that found in the India, Japan and China's forests. The silk provided by willed spp e.i. Eri, Muga, Tussah and Yamamai are heavier and dark in color hence they are less valued than that of Bombyx mori. Silk can be dyed, spun, in to thread and woven in to fabric. Cloth of Silk is warm in winters, cool in summers, light in weight, and resistant to wrinkling.

4.5.2. Production of Shellac

Laccifer lacca, is a scale insect that secret a hard encrustation over the body as a protective covering. It is of brown color usually and these insects grow on acacia trees in India and Burma. Scale insects present on twigs are heated to extract the resins and then purify. One gram of Lac is extracted from Up to 200 insects. In present the synthetic material such as Polyurethane and vinyl has been taken place of Lac, even after Lac is still in use as dyes, inks, polishes, sealing waxes, and as stiffening agents in the fabrication of felt hats. It is animal originated and commercial resin.

4.5.3 Production of Cochineal

Cochineal pigments use in Painting: A scale insect Dacylopius coccus found in Mexico and Central America on prickly pear cacti. Cochineal pigment is extracted from these scale insects. For the first time it was used by Aztec Indians as medicines, body paints and as textile dye. The cochineal pigment was important for the intensity and permanency of colors. It was very costly because of its scarcity, so it was used in only the finest fabrics. Now a day's aniline dyes have taken place of Cochineal in textile industries which is very economic. But the cochineal pigment is still giving the colors in foods, beverages, cosmetics (lipsticks) and art product.

Table 1: Natural enemies and their use

Predator. Parantod	I :rap	Beneficial insect or lincrlebrate	Pat attacked	Impact as pest
1.'"blun	Beetles Pawner)	I adyln-ds I Fund, Conn:OW.1a Red and Blue beetles (Daanolsnal bellalusi Green crab beetles (Canna *sawn). Caen solar beetles ICAmiliognala panel/AO	Aphids. mates. aim. meal,ags. moth eggs including ifehorhar spp and larvae.	Able to handle a wide range of prey and art minantely effective. Some specks fag. ladybirds I heath the ark and lax are predawnn.
Predators	Bugs Menupteral	Assam bugs (Family Redutudick Bigeyed bap (Getnern &brat brown smudge bugs toenattvotti inmates). Damsel bugs War barberry, i. glossy add bug (Cennaralan audits I. Pirate bug I Onus pp 1. Apple dimple bug (Glyn bunny) hetilnean I. tinned reason shield bug I On haha A Broken backed hue I hts / only es s pall damn	Aphid D.diwk math eggs of and lax or !retina Is pp. CillaCTIS. agniltiotera 1:torah false Dopers	Picas pest using mouthNn• and then sucks out interior Depending on the species a predatorybug. adults. law or eggs ma, a hesnackedsnackedked

Predator. Parantod	I :rap	Beneficial insect or lincrlebrate	Pat attacked	Impact as pest
Predates	Predacity lane	Moonily Laic I Family Syryban	Aphads	lava spear anal, with ows al suskiwi internaljuices. Adult hoserfly arc at reeslacan
Pr.-dabs+	'Nees 1 Acanna I	Predatory mates from del fcrent Families. e.g. Assystulae.Bdcllidae, linthracida, Pennlnlac and Canova	Blue (tat ale. Linea ilea, Red. kern, rah ate	Predanous on other mitemorns and facets fleasI Sanaa troth. i
Predators	Laceangs	Greco (Against,. it gnaw I and brown Wonsan talmatuar I Lacewings	Aphids. tooth tanacand ergs. v.horny. slain. mites and grab hugs	Larvae tan taws into aftbodies insect. and eggs and suck out contents Lariat of both Bran and green lannags arc predawn! Adult bran !axing, In on hcbotta tees and mites
Predators	Sprier.	Nanny of spines asluding nullsgsders. mghtstalking widen. othvaven, tangleseats spiders. float spann prinpn so en and Inn sat der.	Predators or a range of erns k pests	Pt" species me <0.,unied
	Aphid Parana&	Pat...tont. r Complananti, .4phnhut en t. tv mph/thus frank "pct. Annaba, ...natant	Aphids	Wasp nuns egg Into aplud The developing La ae <seaman, killing the an *mumm,* .t. the adult nap earns
Parana&	Canaria Pampas	Hymenoptera Numerous p.tfl,111: Vise including Banded rascally pause Its hnettannt prom:stoniest. Two toned carpala parasite I ///// ',Telma Y. apogean) i I: aid, kheamonam. Mu oplins √rasohnir. (arena pp daily lirwonidae i	Miran and Whet 1001b 1711C	Female lays eggs in host pupae as the partutand larvae &scion in the hint it canes the death of the Pupa
Panama.	Caterpillar Pausal,	Sorghum nudge parasite. truprlanis ..agiratienos. Aperohn clan √ap(asass. retrain, ht et sop i	Ynn !him idge	Wasp lays egg-, in nudge larvae and emerges al pupal mane.
Parana	Catragallir Patna!,	Tachund flies	Delman. lager. armystrak Fra`kbutPur and °dr: larvae	Female lays eggs in hat pupae as the pantos.] larvae develop in the host it c s the death of the PUP.

Parasitoids	Ilelicoverpa Egg parasitoids	Hymenoptera: Trichogrammtr (Family Trichogrammatidae) and Telenomus (Family Scelionidael egg parasitoids	Hehcoverpa and other Lepidoptera	Tiny wasps that parasitise Lepidopteran
Parasitoids	Whitey Parasitoids	Ereanocerus spp. and Encarsia spp. including Encarsia Formosa	Whitefly	Small parasitoid wasps that attack u hastily nymphs.
Patasitoids	GVB egg Parasitoids	Triode= basalis	Green vegetable bug	Small black wasp that parasitises GVB: dotsn't distinguish between eggs of pests and beneficial. and will also parasitise eggs of predatory shield bugs.

Table 2: Weed killers

Weed	Scientific name	Biotic agent /Insects
Prickly pear	Opuntia dilleni	Dactylopius opuntiee
Congress grass or Carrot weed	Parihrnium hysterophono	Zygogramma bicolorata
Lantana weed	Lantana cumin,	Ophiomyia lantanae
Siam weed	Choromoforna &tram	Pariuchaetes pseudoinsulata
Water fern	Salvinia molest('	Cryptobagus singulars

4.5.4 Production of Tannic Acid

Tannic acid first produced by an abnormal pant growth found on oak trees in Asia known as Allepo gall. Tiny wasps (Family Cynipidae) secrete some chemical and in response of it the tree produces gall tissues. Tannic acid is a chemical compound used in the dying, in leather industries, for tanning and in the manufacture of some inks. It can also be extracted economically from Quebracho tree, hence there is no commercial market for oak gall is present today.

4.5.5 Insects as Food

Human ancestors were used to get nutrition from Insects. Even today, the insects are being used by people as food in many countries. High in protein and low in fat dried grasshopper are sold in village markets of Mexico. Insects are mixed with flour to make tortillas and can be fried or ground into meal. Wood-boring beetle's larvae can be boiled or roasted over a fire. And there is long list of nutritive edible insects e.g. Ants, bees, termites, water grubs, caterpillars, flies, crickets, katydids, beetle larvae, and nymphs of dragon fly are among the list. And in Thailand the pupa of silkworm is used as food for human being.

4.5.6 Insects as Medicines

Since ancient times Insects derived products have been widely used in medicines (table-3). Maggots and honey showed healing property in chronic and post-surgical wounds. And honey is also being used to treat burns and combining with bee wax it found curative for the dermatological disorders. Another product of honey is royal jelly is used to treat post-menopausal symptoms. A derivative of blister beetle Cantharidin being used in treating. **Skin** dionsos'.

Table-3 some Insects ono tneir prooucts as memcme

Insects / insect products	Uses
Maggots	Wounds Healing
Honey	Wounds Healing, skin disease. infection
Royal-jelly	Post Menopausal symptoms
Been and Ant venom	Joints pain
Propolis	Infection
Canthiridine	Skin ThintieS

4.6 GARDEN ATTRACTS MANY BENEFICIAL INSECTS

The larva of many beneficial insect predators e.g. Predators such as hover flies, lacewings, lady beetle, and parasitic wasps feed on large number of harmful insect pests. But adult of these feed on pollen and nectar. Provide adults with food and habitat in your garden with a variety of plants will encourage the production of all life stages of insect predators and will help in reducing undesirable insect pests naturally.

Variety of plantings can be create to provide habitat and food for different insect species and life stages- eggs, larvae, pupa and adults. Many small flower plants including yarrow, sunflower, alyssum, asters, cosmos, mints, lobelia, basil, stonecrops, thyme, parsley, dill, borage and many other herbs are preferable. In garden eco-system natural predator consume their food from pests present. Pesticides kill many beneficial insects. So, Eliminating use of pesticide encourage beneficial insects.

Variety of plantings can be created to provide habitat and food for different insect species and life stages- eggs, larvae, pupa and adults. Many small flower plants including yarrow, sunflower, alyssum, asters, cosmos, mints, lobelia, basil, stonecrops, thyme, parsley, dill, borage and many other herbs are preferable. In garden eco-system natural predator consume their food from pests present. Pesticides kill many beneficial insects. So Eliminating use of pesticide encourage beneficial insects.

Fig. 1: Some beneficial conclusion

4.8 BENEFICIAL INSECTS: AT A GLANCE

Above description shows the economic importance and use of beneficial insects. Lack of awareness among the farmers about these insects and their benefits they use different management practices to kill them along with the insect-pests. We should have broad principles for supporting the beneficial insects and more specific management practices. Training and program should be run for the awareness among the farmers about the beneficial insects and activities likely to harm them, limited use of broad-spectrum insecticides, habitat conservation in the form of larger patches of remnant vegetation for their survival. Better understanding of the benefits can make conservation more effective and more harmonious land use with effective crop production.

5

METHODS OF BEEKEEPING

5.1 INTRODUCTION

Beekeeping is an art and skill maintaining the bees in modern movable frame hives for hobby or fascination, production of hive products (honey, bee wax etc.) and for pollination services OR the practice of rearing bee is called beekeeping or Apiculture. Apiculture is synonym of the beekeeping and is derived from Latin word *'Apiscultura'*. Apis means 'bee' and *cultura* means 'cultivation through education'. The place where the hives are maintained is called an Apiary. Beekeeping is a high profit enterprise it can be taken up both as subsidiary industry and as well as whole time profession.

Initially in 1953 as many as 230 beekeepers, who maintained around 800 bee colonies in modern bee boxes and were producing 1, 200 Kg of honey annually. Presently it is estimated that with 25.00 Lakhs of bee colonies, 2.50 Lakhs beekeepers and wild honey collectors' harvest around 56, 579 MT of honey in country, which valued Rs. 476.04 crores. The average annual per capita consumption in India is 8.4 g.

5.2 HISTORY OF BEEKEEPING –WORLD

It is not clear when man started beekeeping, but there are archaeological evidences that about 4,000 years ago, the Egyptians kept bees in clay pots and used not only for honey, but also for propolis and wax. In fact, the honeybee was the symbol of Lower Egypt. Still many rock and cave paintings are available across the world depicting the honey bee in different shapes. In ancient Greece and Rome, apiculture was a common practice. The philosopher Aristotle in his book "Historia Animalum" talked about honeybees' floral fidelity, division of labour within the colony and winter feeding. He also described some brood disease. Hippocrates, the Father of Medicine, depicts the nutritional and pharmaceutical value of honey. Greek athletes used honey as an energy boost.

Commercial beekeeping started during the second half of the 19th century. In 1851, Rev. L. L. Langstroth discovered the concept of 'bee space' (3/8 inch space is kept by the bees between two adjacent combs as their passage for free movement all around the combs). Bee space or passage way is the space required between any two frames for the bees to move about conveniently between two combs. Based on this concept, modern age 'Langstroth bee hive' with movable parallel frames/combs was developed by L. L. Langstroth is known as Father of Modern Beekeeping.

5.3 HISTORY OF BEEKEEPING – INDIA

Bees and honey were known to human being in India since time immemorial as their references are mentioned in epics, on murals, sculptures, etc. Vaishali Stupas in Muzaffarpur (Bihar) were built in commemoration of offering of honey to Lord Buddha by king of monkeys and his people whenever Lord Buddha visited the place. Several references of bees have been made in the oldest scripture of India, the Rig Veda.

The earliest method of keeping bees was to use hollowed out tree trucks, empty pots or any other suitable receptacles smeared with wax and sweet scented leaves of Cinnamomum iners on the inner surface; these receptacles were kept in jungles to entice (invite) the bees during swarming seasons. When the bees had settled there, these receptacles were carried to and kept in desired places. This type of hive is called pot hive and it was in practice in Mysore, Coorg, Malabar, Godavari, Kasmir, etc.

In our country, first attempt to keep honey bees in movable frame hive was made in early 1880s in pre-partition Bengal and Punjab. Commercial beekeeping in India started in 1910 in South when Rev. Newton devised a movable frame hive suitable for Asiatic hive bee, Apis cerana. This hive was named after him as 'Newton Hive'. This hive is still popular for keeping the indigenous hive bee, A. cerana. During 1911-17, Newton also trained a large number of beekeepers in Southern India.

The Royal Commission on Agriculture (1928) recommended development of beekeeping as a cottage industry in India. The All India Beekeepers' Association (AIBA) was established in 1938-39. This association started publishing the Indian Bee Journal (IBJ). During 1880, high yielding European bees, A. mellifera, were introduced in our country. A sizable quantity of this species was imported from 1920 to 1951 in the states of Maharashtra, Kerala, Karnataka, Tamil Nadu, West Bengal, Punjab and Kashmir but none succeeded to establish this exotic honey bee species in the country.

5.4 STRENGTHENING OF BEE KEEPING RESEARCH AND DEVLOLNMENT IN THE COUNTRY

After independence, Khadi and Village Industries Commission (KVIC), Govt. of India took up beekeeping as one of its ventures. Some states like Jammu and Kashmir, Karnataka, Uttar Pradesh and Himachal Pradesh established Departments of Beekeeping under their Ministry of Agriculture/Industries. Further, considering the importance of applied and basic research in apiculture, KVIC established Central Bee Research and Training Institute (CBRTI) at Pune (Maharashtra) in 1962.

The research in beekeeping started when Indian Council of Agricultural Research (ICAR), New Delhi started funding to the different projects. Two Beekeeping Research Stations were also established at Nagrota-Bagwan (erstwhile Punjab, now in H.P.) in 1945 and at Coimbatore (Tamil Nadu) in 1951. Recently in Gujarat, Department on Entomology, N.M. College of Agriculture, Navsari Agricultural University, Navsari has initiated research on honey bees and other pollinators sponsored through ICAR, New Delhi with Project entitled "All India Coordinated Research Project on Honey bees and Pollinators from the year 2015-16"

Successful Introduction and Establishment of *Apis mellifera* IN INDIA

After a long gap of unsuccessful attempts of A. mellifera introduction in our country, Professor A. S. Atwal, an Entomologist of the Punjab Agricultural University (PAU), Ludhiana with his associates introduced A. mellifera in 1962 at Beekeeping Research Stations of Nagrota-Bagwan (H.P.) by adopting the 'Inter-specific Queen Introduction Technique'. They imported disease free A. mellifera gravid queens along with worker bees. Later the worker bees were burnt and A. mellifera queens were introduced one each into the de-queened colonies of Asiatic hive bee (A. cerana). After the adaptation of A. mellifera queens, the workers of Asiatic hive bee (A. cerana) reared the brood. It resulted in gradual replacement of workers of A. cerana who died with the age. Thus, A. mellifera stocks were further strengthened by importing disease free consignments of the gravid queen bees.

Convinced with the performance of A. mellifera in the Punjab, H.P. and Haryana and due to the outbreak of Thai Sac brood Viral Disease causing large scale mortality of A. cerana colonies during late 1970s to early 1980s in the states, practicing A. cerana beekeepers of many other states expressed desire to adopt A. mellifera. Due to this, ICAR in 1986 decided to extend this species from Punjab to other states. Now, this exotic honey bee (A. mellifera) has been spread to almost whole of the country. During 1993, Department of Agriculture and Cooperation (DAC), Ministry of Agriculture, Govt. of India laid special emphasis on beekeeping and started a National Scheme on the 'Development of Beekeeping for Increasing Crop Productivity'. Under this Scheme, beekeeping research, training and development projects were sanctioned to various State Agriculture Universities (SAUs), State Agriculture Departments, Government and Non-Government organizations (NGOs). Govt. of India established National Bee Board in 2006.

1. Honey bee:

Scientific classification		
Kingdom	:	Animalia
Phylum	:	Arthropoda
Class	:	Insecta
Order	:	Hymenoptera
Family	:	Apidae
Subfamily	:	Apinae
Tribe	:	Apini
		Latreille, 1802
Genus	:	Apis
		Linnaeus, 1758

5.5 BIOLOGY/ LIFE HISTORY OF HONEY BEE

5.5.1 Eggs

Eggs are laid by queen and when a colony wants to produces a new queen, the special cell is constructed at the lower border of the brood comb. On these cells, singles egg is laid by the queen in each cell which hatched after 3 days. The newly hatched grubs are provided with royal jelly. The grub is fully developed in 5 or 6 days and then queen cell is capped where grub changes into pupa and after a week adults come out by biting the cap of queen cell. The adult who comes out earlier become the daughter queen and it kills the remaining pupae before their emergence.

5.5.2 Nuptial Flight

After 2-3 days the daughter queen takes nuptial flight accompanied by hundreds of drones during day. She overtakes drone in flight. The drone which follows her takes the chance of copulation. The male soon dies after copulation and the mated queen return to the comb. She mate only once in her life time. The seminal fluid (male sperms) is collected in a special receptacle (spermatheca) and used as and when required.

5.5.3 Drone Honeybee Die After Copulation.......WHY?

The drone mounts the queen, inserts his endophallus, and ejaculates his semen. During ejaculation, the male falls back and his endophallus is ripped out of his body and remains attached to the queen. Drones mounting later remove the previous drone's endophallus and lose their own through similar matings. The emasculated drones die very quickly with their abdomens burst in this fashion.

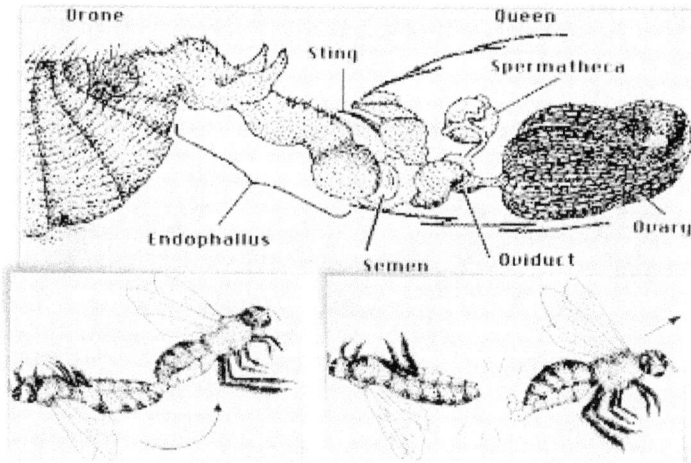

5.5.4 Oviposition

After some times the daughter queen starts eggs laying and is called as mother queen. She lays fertilized or unfertilized eggs at her will. Once egg is laid in a cell, it hatched in 3-4 days. The eggs are long, oval and light brown in colour.

The queen measures the cell opening with her front legs as she inspects each cell prior to laying her egg. Worker bees develop horizontally in hexagonal cells of approximately 0.2 inch (5 mm) diameter (5 cells/inch). Drones develop in slightly larger horizontal cells. The female queen develops in a vertically-oriented cell. The existing queen herself lays fertilized eggs in special cup-like structures, called queen cups, oriented vertically on the face of the horizontal worker and drone comb or more usually at the bottom margin of the comb.

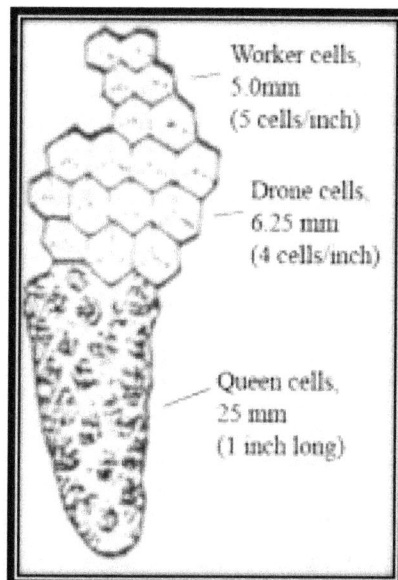

Life cycle of honeybees

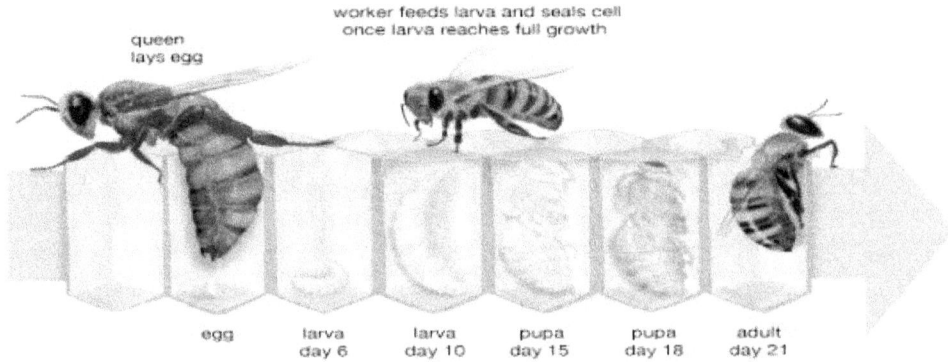

worker feeds larva and seals cell
once larva reaches full growth

queen
lays egg

| egg | larva day 6 | larva day 10 | pupa day 15 | pupa day 18 | adult day 21 |

5.5.5 Grub

From the fertilized eggs the queen and worker are developed, while from the unfertilized eggs drones are developed. The grubs are cylindrical in shape and light yellow in colour, they fed with the royal jelly for 2-3 days after that they are provided with honey and nectar, etc. The grub period lasts for about 5-6 days. Worker bees are raised in the multipurpose, horizontally arranged cells of the comb. Future workers receive the royal jelly only during the first 3 days as compared to future queens, who are fed with royal jelly throughout their larval life. The developing queen larva is always surrounded by royal jelly, a special highly nutritious food produced by head glands of the workers. This feeding scheme, called massive provisioning is unique to the queen and continues throughout her entire developmental period. Worker bees mix the honey with pollen and feed to the drone larvae. Future drones receive royal jelly for the first 3 days. After that, they are shifted to progressive feeding as discussed in worker feeding.

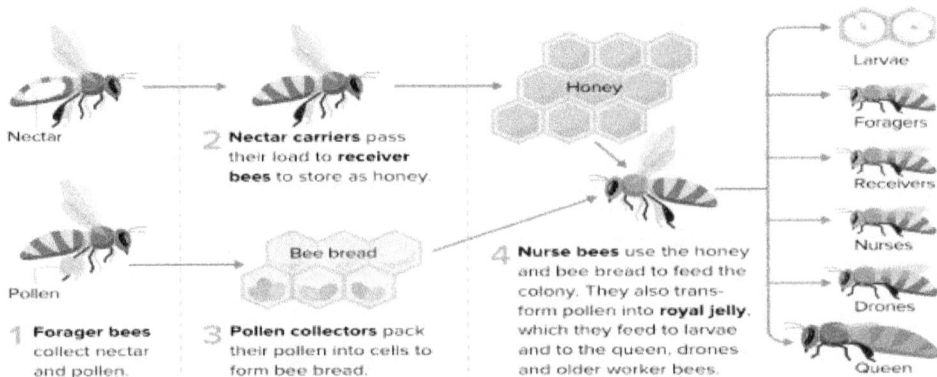

Nectar

2 **Nectar carriers** pass their load to **receiver bees** to store as honey.

Honey

Larvae

Foragers

Receivers

Nurses

Pollen

Bee bread

1 **Forager bees** collect nectar and pollen.

3 **Pollen collectors** pack their pollen into cells to form bee bread.

4 **Nurse bees** use the honey and bee bread to feed the colony. They also transform pollen into **royal jelly**, which they feed to larvae and to the queen, drones and older worker bees.

Drones

Queen

5.5.6 Pupa

Full grown grub forms a cocoon and pupates inside the cell. The pupal periods lasts for about 7-14 days depending upon the type of adult to be produced. The time required for development of different castes of A. mellifera is given below:

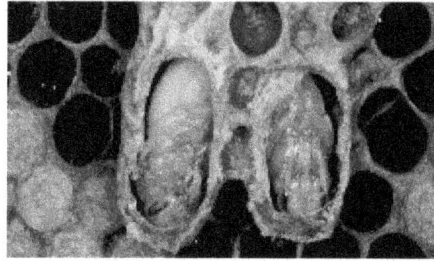

Adult	Eggs	Grub	Pupa	Total
Queen	3 days	6.5 days	6.5 days	16 days
Worker	3 days	8.0 days	10.0 days	21 days
Drone	3 days	9.5 days	11.5 days	24 days

5.5.7 Total lifespan/biology of honey bee

Development Stage	Castes		
	Queen	Worker	Drone
Egg	3 days	3 days	3 days
Unsealed stage	5 days	5 days	7 days
Cell scaled	8th day	8th day	10th day
Cocoon information	10th day	Ilth day	14th day
Adult formation	15th day	20th day	22" day
Adult emergence	16th day	21" day	24th day
Sexual maturity	Within 2-3 days	-	13 days
Adult longevity	3-4 years	6 weeks	2-3 months

Honey bees are the only bees to die after stinging ....WHY??

Because when a honey bee stings a person, it cannot pull the barbed stinger back out. It leaves behind not only the stinger, but also part of its abdomen and digestive tract, plus muscles and nerves. This massive abdominal rupture kills the honey bee.

5.5.8 Honey Bee Castes

The honeybee is a social insect and lives in colonies with a highly organized system of division of labour. Many combs are found in a colony in which the members of the same family used to live. Each family consists of three castes: queen (fertile female), drones (males) and workers (sterile females). Each caste has its special function in the colony. The workers are undeveloped females, the drones are known as males and the queen is the fully developed female. Every honey bee colony comprises of 35000 to 70,000 members includes a queen, 200-300 drones and several thousand workers.

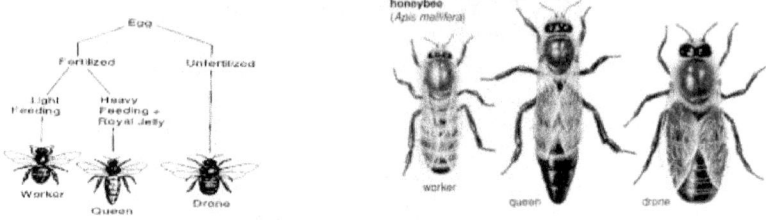

Sex differentiation in bees

Queen: The queen is a true mother bee. Queen is the only female that is completely developed sexually from fertilized egg. This is a result of a total diet of royal jelly during the developmental period. She has a long abdomen extending well beyond the apical margins of the wings. In the colony, she is found in the area of the brood nest. A well developed queen is generally two to three times bigger than a worker and measures about 15-20 mm in length.

Duties of a queen

1. The only individual which lays eggs in a colony (Mother of all bees).
2. Lays upto 2000 eggs/day in Apis mellifera and maintaining a populous colony.
3. Five to ten days after emergence, she mates with drones in one or more nuptial flights.
4. When her spermatheca is filled with sperms, she will start laying eggs and will not mate any more.
5. She lives for 3 years and when it is weak or unable to lay eggs it is replaced by one of the daughter queen.
6. The secretion from mandibular gland of the queen is called queen's substance.
7. The queen substance if present in sufficient quantity performs following functions
   a. Prevent swarming and absconding of colonies.
   b. Prevent development of ovary in workers
   c. Colony cohesion is maintained.
8. 8. The queen can lay either fertilized or sterile eggs depending on the requirement.

5.5.9 Drone

Drones, the functional males of the colony are produced from unfertilized eggs, and are larger and darker than the worker. It is smaller than queen and measures about 15-17 mm in length. Drones are not a permanent member of colony. The queen can control whether or not the egg is fertilized as she lays it. The compound eyes are holoptic i.e. very large and are united at the vertex. The end of the abdomen is blunt and is covered with a tuft of small hairs. Drones cannot sting. As the sting is a modified structure of the female genitalia, drones do not have stings. They also do not have any of the structures necessary to collect nectar and pollen. It dies after successful mating with the queen.

Duties of a drone
1. Their important duty is to fertilize the queen.
2. They also help in maintenance of hive temperature.
3. They cannot collect nectar / pollen and they do not possess a sting.

5.5.10 Workers

Workers are sexually sterile female caste and is the smallest in size as compared with the above two castes. On ventral side of the abdomen, wax glands are present. Hind legs are modified for pollen collection. The mandibles are flattened and spoon shaped which are used for molding the wax for comb building. They do the work of the colony and maintain it in good condition. Workers have special structures and organs which are associated with the duties they perform.

Duties of a worker
1. Their adult life span of around 6 weeks can be divided into:
   ii. First three weeks- house hold duty.
   iii. Rest of the life- out door duty.

[i] House hold duty includes
a. Build comb with wax secretion from wax glands.
b. Feed the young larvae with royal jelly secreted from hypopharyngeal g land.
c. Feed older larvae with bee-bread (pollen+ honey).
d. Feeding and attending queen.
e. Feeding drones.
f. Cleaning, ventilating and cooling the hive.
g. Guarding the hive.
h. Evaporating nectar and storing honey.

[ii] Outdoor duty includes
1. Collecting nectar, pollen, propolis and water.
2. Ripening honey in honey stomach.

Schedule of a worker bee in the hive

Days after emergence	Task
1-2	Clean cells and warm the brood nest
3-5	Feed older larvae with honey and pollen
6-10	Feed younger larvae with products of the head glands
11-18	Ripen nectar, produce wax and construct comb
19-21	Guard and ventilate the hive, take exercise and orientation flights to learn to fly and locate the hive
22 +	Forage for nectar, pollen, water or propolis

Morphological differentiation among different castes of A. mellifera

Character	Queen bee	Drone bee	Worker bee
Adult			
Body size	Longest	Medium	Smallest
Wings	Do not completely cover the abdomen	Completely cover the abdomen	Completely cover the abdomen
Head	Triangular and little roundish	Roundish	Triangular
Abdomen	Long, gradually tapering	Tip of abdomen blunt and hairy	Tip of abdomen conical and subtly pointed
Compound eyes	Small and well apart	Large kidney shaped, meeting at vertex	Small and well apart
Pollen collecting legs	Not developed	Not developed	Well developed
Sting	Present but without barbs	Absent	Present with barbs

5.6 DIFFERENT SPECIES OF BEES AND THEIR IMPORTANT CHARACTERS:

The honey bees belong to superfamily Apoidea and the family Apidae and the order Hymenoptera. There are six species of Apis viz., Apis cerana, Apis florea, Apis dorsata, A. andreniformis, A. laboriosa and A. koschevnikovi which are indigenous to India and A. mellifera which has been introduced from European countries. The commercialized honey bees in India are two domesticated/hive bees, Apis mellifera Linn. and Apis cerana F. and two well-known wild species, Apis dorsata F. and Apis florea F. They have well developed stings. The Dammer bee or little bee, Trigona iridipennis (Meliponinae) has only a vestigial sting. All five species are social insects living in colonies with remarkable degree of social instincts and division of labour among the different members of the colony.

There are five important species of honey bees as follows.

Scientific name	Common name
Apis dorsata	The rock bee
Apis cerana indica	The Indian hive bee
Apis florea	The little bee
Apis mellifera	The European or Italian bee
Trigona iridipennis (T. laeviceps): Dammer bee, stingless bee	Dammer bee, stingless bee

⦿ The rock bee or giant bee, Apis dorsata Fabricious

  1. It is largest of the honey bees and measuring about 20 mm in length.
  2. It construct single comb of huge size in open (About a meter in diameter)

3. The comb is fully exposed and hung from inaccessible branches of trees, along sides of steep rocks in the forest and even from the walls, rafters and other parts of buildings.

4. It produces plenty of honey i.e. 37 Kg honey /comb/year.

5. It represents a major portion of honey sold in our markets.

6. Rock bees are irritable and ferocious in nature and difficult to rear.

7. They shift the place of the colony often. In winter, they migrate to plains and come back to hills during summer season.

⦿ Indian hive bee/Asian bee, Apis cerana indica Fabricious

1. It is common Indian bee found in both forest as well as in plains throughout country.

2. It is smaller than the rock bee but the larger than the little bee. Bee measures about 15 mm in length. 3. They make multiple parallel combs on trees, cavities, caves in darkness and such other hidden sites, the combs being parallel to the direction of the entrance in the plains and the right angle to the entrance in cold regions.

3. It is mild and capable of being domesticated and is commonly reared in south India.

4. They produce about 2 - 5 Kg of honey/year/colony.

5. A queen can lay 350 – 1000 eggs per day. 7. They are more prone to swarming and absconding. 8. They are native of India/Asia.

⦿ The little bee, Apis florea Fabricious

1. It is known as the little bee since it is smallest of the four species of Apis. Bee measures about 7 mm in length.

2. It is seen only in the plains and not in hills above 450 mt MSL.

3. It does not like darkness therefore forms its comb in the open place e.g. bushes, hedges, buildings, caves, empty cases etc.

4. It builds a single comb which is very small and produces about 0.5 to 1 kg honey/year/hive and so it is not domesticated and reared.

5. A queen can lay 323 – 365 eggs per day.

6. They are not rearable as they frequently change their place.

⦿ European bee or Italian bee, Apis mellifera Linnaeus

1. It is extensively reared in Europe and America.

2. It was introduced in India in the year 1962 by Prof. A. S. Atwal in Nagrota (HP) from European countries (Italy). He is called as "Father of Modern Beekeeping in India".

3. The behaviour and appearance of A. mellifera is similar to A. cerana.

4. It makes its nest in enclosed space (in darkness) in multiple parallel combs and is endowed with all the good qualities of a hive bee, i.e. has a prolific queen, swarms less, gentle tempered so, domesticable, good honey gatherers and can guard its nest against enemies.

5. They yield on an average 45-180Kg honey/hive/year

6. They are larger than Indian bees but smaller than Rock bees.

◉ Dammer bee or stingless bee, Trigona irridipennis Dal. (T. laeviceps)

1. This is the smallest species and differs from other bees in its appearance and habitats.

2. They do not have sting i.e., stingless.

3. They built their comb in hollow walls or tree trunks.

4. They construct their comb with a dark material called "Cerumen" which is a mixture of earth and wax or resin collected from plants as they do not secrete wax to build combs.

5. It is very poor honey gatherers and yields only 60-180 ml/colony/year.

6. Its honey is used in Ayurvedic medicine

5.7 IDENTIFICATION/DIFFERENTIATION AMONG DIFFERENT BEE SPECIES

Sr. No.	Characteristics	The rock Bee, Apis dorsata (Giant honey bee)	The little bee, Apis florea	Indian hive Bee/Asian bee, Apis cerana indica (Asiatic honey bee)	European bee or Italian bee, Apis mellifera Linnaeus	Dammer bee or stingless bee
I.	Body size	Largest	Smallest	Medium	Medium	Smallest
2.	Body colour	Head Blackish, abdomen reddish- yellow anteriorly & black at the tip	Abdomen Orange anteriorly with black & White stripes posteriorly	Body colour Blackish, abdomen with White & black Stripes	Body golden Yellow, profusely hairs With faint black and yellowish stripes posteriorly	Body is Reddish brown in colour
3	Wings	Smoky	Transparent	Transparent	Transparent	Transpare nt
4	Proboscis size	Largest	Smallest	Medium	Medium	Smallest
5	No. of worker cells/4 linear inches	18.75	32.8 to 36.0	21.25 to 25.00	19.3	--
6	Nature and Temperament	Wild bee, hostile	Wild bee, relatively less hostile	Can be hived. docile	Can be hived, docile	Can be hived, docile, do not have sting i.e.. stingless.

5.8 MAJOR ACTIVITIES OF HONEY BEES

The honey bees remain active generally throughout the year except during severe winter. Following are the main activities of honey bees:

5.8 1. Foraging

The field bees get activated in the morning and go out on foraging and collect pollen, nectar, propolis and water, carry them to the hive and make a number of trips till sunset. The bees that go out first to find out new sources of these materials are called searcher bees or scout bees. They return to the hive and communicate the message to young foraging bees by means of definite patterns of dancing. At any time bees collect most of the materials from a single or a few plant species but bees in two different colonies located side by side may visit entirely different sources, mainly due to the differences in discoveries by the scout bees. The bees collect materials from a source till they are exhausted when they may go in search of new areas. The honey bees usually forage within about 100 meters distance from the hive but they can go up to 1.5 km. They are capable of flying at a speed of 25-30 km per hour. The bees are most active in foraging within a temperature range of 25-27°C. The bees do not go out for foraging at wind speed of more than 24 km per hour.

Nectar is collected by the foragers from the flowers and is stored in the crop where it is mixed with saliva. The invertase contained in the saliva acts upon sucrose of the nectar and converts it to dextrose or levulose. The bee returns to the hive and regurgitates the contents of the stomach into comb cells which are covered by flat airtight cappings. The weight of nectar load varies from 25 to 40 mg. On a given trip, a bee visits and exploits 1-500 flowers and makes, on an average, 10-15 trips in a day. During honey flow when there is abundance of food available, bees work to their full capacity and may make upto 150 trips a day. The pollen is collected and carried to the hive by the bees in the pollen baskets located in their hind tibiae. The bee returns to the hive and the pollen pellets are pushed down to the appropriate cells by means of spine in the middle leg. The weight of pollen load varies from 10 to 30 mg. The workers make about 6,000 trips to collect 0.5-1 g of pollen. The propolis is also carried in the pollen basket by the worker bees. As soon as the collecting bee returns to the hive, another worker unloads the propolis from the former, carries the same in its mandibles to the place requiring cementing and presses it into the crevices in the comb.

5.8 .2 Combing

The comb of honey bees comprises of several hexagonal cells on both side of mid-rib. The combs are built with beeswax which is secreted by 4 pairs of wax glands located on 3-6 abdominal sterna. The wax secreted in a liquid form, collects in the intersegmental regions, hardens into thin flakes that are picked up by the legs and passed on to the spatulate mandibles for being kneaded and stuck to the top of nesting cavity and extended downwards bit by bit. Several bees hang like a sting to do the job. Usually, the cells meant for honey storage are located uppermost near the point of attachment below which are pollen cells spread in 5 cm wide band, further down are worker brood cells which are followed by the drone and queen cells. The worker cells are the smallest, drone cells larger than the worker cells and queen cells the largest. Worker and drone cells are directed sideways and queen cells vertically with open ends downwards. Cells of the size of worker and drone cells are used for storing honey and pollen. Cells containing unripe honey or developing brood are uncapped; those with fully ripe honey and fully fed grubs are capped, and pollen cells are generally not capped. Freshly built comb is generally white, but becomes dark after some time.

5.8. 3. Swarming

Swarming is a method of reproduction in which a part of the colony migrates to a new site to make a new colony. During spring and summer when conditions are favourable and food is available in plenty, the bees multiply greatly with the result the comb becomes crowded and the bees begin to make preparations for swarming. At this stage, the daughter queen cells are built at the bottom and when new queen is ready to emerge out, the new queen and a large number of workers which have previously filled the cells with honey, leave the nest to start a new colony. Swarm settles in a suitable place already searched out by the workers for building new comb. In a parent colony the first daughter queen which emerges after swarming, kills the baby queen in the other cell and establishes herself as a mother queen. After that, they start their routine work of gathering nectar and pollen.

5.8.4. Absconding and Migration

Complete desertion of a hive is known as absconding. This may occur due to lack of water, exhaustion of food store (either due to short supply of nectar or robbery of honey), unfavourable environment, constant pest attack (ants, wax moth, etc.) and even by excessive interference by the beekeeper in which case he is regarded as an enemy. Prior to absconding, the bees 'drink' whatever honey their nest has and then migrate leaving behind empty combs, brood and sometimes even food. Absconding can be prevented by providing water or sugar solution near the hive particularly during summer.

5.8.5. Language of Bees

Honey bees have a unique and one of the best understood animal languages with which they inform each other the distance and direction of the source of food. The forager bee on return to the nest makes two kinds of dances on the surface of the comb, i.e. round dance and tailwagging or figure of eight dance (Fig. 1), which the insiders perceive by contacting the forager's body with their antennae. In the round dance, which is used to indicate a short distance (less than 50 m in case of A. mellifera), the bee runs in circles, first in one and then in opposite direction (clock and anticlockwise), while in the tail-wagging dance which is used to indicate a longer distance (beyond 50 m in case of A. mellifera), the bee makes two half-circles in opposite directions with a straight run in between. During the straight run, the bee shakes its abdomen from side to side and the number of wags per unit time is related to the distance the food was located, i.e. more the wags, nearer is the food. The direction of the food is conveyed by the angle that the dancing bee makes its straight run and top of the hive which is the same as between the direction of the food and direction of the sun. Prof. Karl von Frisch was awarded the Nobel Prize in Physiology and Medicine in 1973 for discovering and interpreting the language of the honey bees in early 1920s. Later on, it was found that honey bees employ both dance and sound in their language.

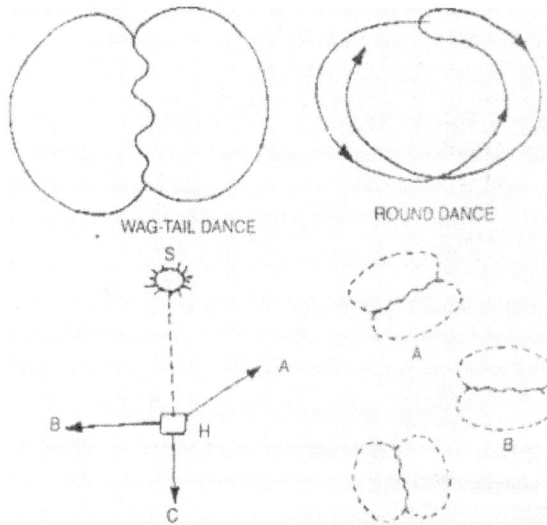

Fig. 1: Communication dances in honey bees according to location of the food source and direction of the sun (A, B, C = Directions. S=Sun. H=Beehive)

5.8.6 6. Air conditioning

Among the living creatures, honey bees are the only organisms which make their comb air-conditioned. They keep their comb warm in winter and cool during summer. The brood temperature is stabilized between 33 and 36°C averaging about 34.5°C. Clustering begins when the temperature inside the nest dips below 18°C and they generate heat by sitting on one another and rubbing their legs due to which the temperature of the comb rises. In summer, when the temperature rises above 33°C, the bees start fanning with their wings at the gate inside as well as outside with the result water evaporates from the honey and comb remains cool. The brood nest is usually kept at 40 per cent relative humidity. A large number of foragers start collecting water from outside, that is received by house bees inside and carried by them to the site where most needed and evaporated They spread minute drops of water in cells and also form thin films from regurgitated water on their tongues for evaporation. In the event of extreme hot weather, they even suspend collecting concentrated nectars but prefer dilute nectars in case water is in scarcity, as the dilute nectars may be used for making thin films.

5.9 COMMERCIAL METHODS OF REARING OF HONEY BEES

Rearing the bees in artificial hives is known as beekeeping or apiculture. In India, the beekeeping industry started with the designing of a small hive suitable for A. cerana by Rev. Father Newton in 1910. This hive named 'Newton Hive' is still popular for keeping of A. cerana. The father also trained a large number of beekeepers in Southern India and helped them to establish beekeeping as an economically viable proposition. Mahatma Gandhi realized the importance of beekeeping and included it in his rural development programmes. He inspired rural freedom fighters to take up beekeeping as a venture of livelihood. In earlier times, people after wrapping the blanket on the body or after smoking in night and collect honey from the comb. This was a crude method. After some times, people thought to keep the honey bees and many villagers took interest in keeping honey bees and provided various types of hives in their houses. Thus, beekeeping can be divided into primitive and modern methods:

5.9.1 Primitive or Indigenous Methods

This is primitive and unplanned method of apiculture. In this method, two types of hives are used,

- **Fixed type:** Providing a receptacle in the wall of the house with an entrance and observation holes.

- **Movable type**: Providing a basket, empty boxes, hollowed logs, bamboo, mud pipes, earthen pots, etc. Anything that can protect bees from sun and rain.

In the indigenous method, the bees are first killed or made to escape from the hive with the help of smoke when the bees are at rest during night. This method has many drawbacks and it is not suitable for commercial large-scale production of honey. The following are the disadvantages of indigenous method:

1. The honey cannot be extracted in the pure form. The extracted honey also contains the larvae, pupae and pollen cells.
2. The future yield of the honey is affected as the colony has to be destroyed to extract the honey. Moreover, it takes lot of energy of the bees to build new hive.
3. The bees may not construct the new hive in the same place as the old one.
4. The natural hives also have the danger of attack by the enemies like rats, monkeys, ants etc. The natural hives can also be damaged by the climatic factors.
5. Also scientific intervention is difficult in the indigenous method and thus improving of the bee race is impossible.

5.9.2 Modern Method or Frame Hive Method

Modern Method or Frame Hive Method Frame hives are fitted with movable frames on which the bees are persuaded to build their combs. They are usually composed of several boxes, one on top of the other, in which hive frames are suspended. The lower boxes (1-2) are used for holding the brood and the upper ones (1-2) are used for collection of honey, pollen and propolis. The artificial comb was first introduced by Revd. L.L. Langstroth in 1851 in America. In India, during 1910 Rev. Father Newton designed a small hive suitable for A. cerana.

5.9.3 Beekeeping with Frame Hive Method

Apiary is the place where the honey bees are reared for honey and wax either commercially or as a hobby. Often a beekeeper is left with no choice for location of his hives, when he intends to keep them in his backyard or a small home garden. But where a selection among many possible sites can be exercised, the following points.

Requirements for site selection for apiary:

- Apiary should be located where there is abundance of nectar and pollen yielding plants within the radius of one to one and half kilometer.
- The site should not be exposed to strong winds or at least the hives should not face the direction of the prevailing winds. Trees and bushes may be provided to make the site less windy.
- The site should be flat but with good drainage facilities.
- Clean and fresh running water should be available to the bees in or near the apiary.
- A young orchard is an ideal choice.
- If the site is shade less and exposed, an artificial shade may be provided.
- An apiary should not be located too near highways.
- A good barbed wire fence or live hedge may be provided to keep out intruder/ thief.
- The site should be free from termite and black ant infestation.

BEE HIVES: Various types of bee hives are available for beekeeping. They are wooden boxes having two parts: upper ¼ comb is chamber and lower ¾ is brood chamber. Following types of bee boxes are used in beekeeping.

Sr. No.	Box type	Dimensions	Remark
I.	Ghos box	36 cmx 21.5 cm	These two types of bee hives popular in India. In more are India Newton's beehive are manufactured based on Bureau of Indian Standards (BIS) specifications and called as BIS hives.
2	Newton box (BIS hive)	20.2 cm x 14.0 cm	
3	Langstroth hive	42.2 cm x 31.1 cm	Some other familiar bee

			Boxes. Nowadays these boxes are widely used in commercial beekeeping. Langstroth hive is suited to A. meth:fern.
	(American hive)		
4	Pant, Kanje and Jeolikote No.I	42.2 cm x 12.3 cm	
5	Dadant box (Russian hive)	47 cm x 28.6 cm	
6	Thompson box	30.5 cm x 15.2 cm	

Hive parameters	BIS hive C type for A. meth:fens (Modified langstroth type)	BIS hive A & B type for A. cerana (Modified Newton and Jeolikote types)
Frames	Contains 10 frames	May contain 4, 8 or 10 frames
Super Chamber	Generally full super chamber is used.	Half (shallow) super chamber is generally used.
Brood/super frame size	Outside: 448 x 232 mm Inside : 428 x 192 mm	Type A: Modified Newton Type Outside: 230 x 165 mm Inside : 210 x 145 mm Type B: Modified Jeolikote Type Outside : 300 x 195 mm Inside : 280 x 175 mm
Bee space	10 imn	Type A : 7 to 9 mm Type B : 8 or 9 mm

5.10 EQUIPMENTS USED IN COMMERCIAL BEEKEEPING

A movable frame hive is composed of the following parts/appliances.

1. **Bee hive:**

   ◉ It is movable wooden hive for bees with an entrance and parallel movable frames on which bees raise their combs.

   ◉ It provides protection to the colony from adverse effects of external environment. The important parts of the hive are bottom/floor board with alighting board, entrance, lower/brood chamber, frames, dummy board, super/honey chamber, inner cover (crown board) and top cover.

2. **Nucleus hive:**

   ◉ Small bee hive for keeping 4-6 frames. These are used for mating of queens and division of colonies.

3. **Observation hive:**

   ◉ Small hive with glass sides to observe movements and behaviour of bees.

4. **Synthetic combs:**

    ◉ It is made up of high density polythene (plastic). It can be used in both super and brood chambers.

    ◉ Since the comb is fully moulded, bees only put wax caps on the cells.

    ◉ Advantages of synthetic combs viz., More honey can be extracted, Combs can be easily sterilized, Resist to wax moth attack, Combs will not be damaged during honey extraction.

5. **Hive stand:**

    ◉ This is used to keep the bee hive above the ground so as to protect the colony from termites, ants and other crawling insects and also prevent soil moisture getting into thehive or facilitate ventilation from below the hive.

    ◉ The stand is made of wood or iron tubing or angle iron.

    ◉ Any four legged stand of 15-25 cm high is sufficient.

    ◉ Ant wells of 15 cm in diameter kept under four legs to prevent ants and other crawling insects entering into the hive.

6. **Bottom board:**

    ◉ It forms the floor of the hive made up of a single piece of wood or two pieces of wood joined together.

    ◉ Wooden beading are fixed on to the lateral sides and back side.

    ◉ There is a removable entrance rod in the front side with two entrance slits to alter the size of the hive entrance based on need.

    ◉ The board is extended by 10 cm in front of the hive body which provides a landing platform for bees.

    ◉ Size of alighting board is 40 x 28 cm (BIS hive).

7. **Brood chamber:**

    ◉ It is a four sided rectangular wooden box without a top and bottom.

    ◉ It is kept on the floor board.

    ◉ A rabbet is cut in the front and back walls of the brood chamber.

    ◉ The brood frames rest on the rabbet walls.

    ◉ In brood frames, bees develop comb to rear brood.

    ◉ Size of brood frame is (outer dimensions) 29 x 29 x 17 cm.

    ◉ There will be 8 frames. Length and height of frame is 20.5 x 14.0 cm (BIS hive).

8. **Super chamber:**
   - ◉ It is kept over the brood chamber and its construction is similar to that of brood chamber.
   - ◉ Super frames are hung inside.
   - ◉ The length and width of this chamber is similar to that of brood chamber.
   - ◉ The height may also be similar if it is full depth super as in Langstroth hive. But the height will be only half if it in a shallow super as in Newton's hive.
   - ◉ Surplus honey is stored in super chamber.

9. **Hive cover/Top cover:**
   - ◉ It insulates the interior of the hive.
   - ◉ In Newton's hive, it has sloping planks on either side.
   - ◉ On the inner ceiling plank there is a square ventilation hole fitted with wire gauze.
   - ◉ Two holes present in the front and rear also help in air circulation.
   - ◉ In Langstroth hive and BIS hive, the hive cover consists of a crown board or inner cover and an outer cover.

10. **Inner cover:**
    - ◉ The inner cover is provided with a central ventilation hole covered with wire gauze help in air circulation.
    - ◉ The outer cover is covered over with a metallic sheet to make it water proof to rain water.

11. **Hive frames:**
    - ◉ The frames are so constructed that a series of them may be placed in a vertical position in the brood chamber or the super chamber so as to leave space in between them for bees to move.
    - ◉ Each frame consists of a top bar, two side bars and a bottom bar nailed together.
    - ◉ Both the ends of the top-bar protrude so that the frame can rest on the rabbet.

12. **Dummy or Division board/ Movable wall:**
    - ◉ It is a wooden board slightly larger than the brood frame.

It is placed inside the brood chamber.

- It prevents the bees from going beyond it.
- It can be used as a movable wall there by limiting the volume of brood chamber which will help the bees to maintain the hive temperature and to protect them from enemies.
- It is useful in managing small colonies.

13. **Bee feeder:**
    - Used for providing sugar syrup as feed to the bees during dearth period.
    - A normal method of providing feeding is to keep a can with small holes punched on its lid. The can is filled with sugar syrup and kept over the frames in an inverted position.

14. **Queen excluder:**
    - It is made up of perforated zinc sheet.
    - The slots are large enough to allow the workers to pass through but too narrow for the queen.
    - A wire grid/dividing grid with parallel wire mounts can also be used as a queen excluder.
    - It is inserted in between the brood frames and super chamber.

15. **Queen gate:**
    - It is a piece of queen excluder sheet and fitted on the slot of entrance gate.
    - The holes in the sheet are large enough to allow free movement of worker bees in and out of the hive, but too small to allow queen's passage.
    - It confines the queen inside the hive. It is useful to prevent swarming and absconding. It also prevents the entry of bee enemies like wasps into the hive.

16. **Queen cage:**
    - This is used for transport of queen either with a few attendant worker bees, in packages.
    - It is a cage made up of wood or wire gauge or plastic structure. This is useful for queen introduction.

17. **Queen cell protector:**
    - It is a cone shaped structure made of a piece of wire wound spirally. It fits around a queen cell.
    - It is used to protect the queen cell, given from a queen right to queen fewer colonies until its acceptance by bees.

18. **Swarm trap:**
    - It is a rectangular box used to trap and carry the swarm.
    - It is fixed near the hive entrance with one or two combs inside during the swarming season.
    - This box traps and retains the queen only. But the swarm coming out from the hive reenters the hive and settles on the comb, since the queen is trapped.

19. **Drone excluder or drone trap:**
    - It is a rectangular box with one side open. The other side is fitted with queen excluder sheet.
    - At the bottom of the box there is a space for movement of worker bees. There are two hollow cones at the bottom wall of the box.
    - Drones entering through the cones into the box get trapped.
    - The narrow end of the cone is wide enough to let the bees pass out but not large enough to attract their attention or re-entry. This device is used at the entrance to reduce the drone population inside the hive.

20. **Pollen trap:**
    - Pollen trapping screen inside this trap scrapes pellets from the legs of the returning foragers.
    - It is set at the hive entrance.
    - The collected pollen pellets fall into a drawer type of receiving tray.

21. **Hive tool:**
    - It is a piece of flattened iron with flattened down edge at one end.
    - It is useful to separate hive parts and frames glued together with propolis.
    - It is also useful in scrapping excess propolis or wax and superfluous combs or wax from various parts of the hive.

22. **Protective dress:**
    (a) Bee veil:
    - It is worn over the face for protection against stings.
    - It is particularly useful for a beginner, for protecting face from bee stings during the handling of bees.

    (b) Gloves:
    - These are used while inspecting and handling colonies to protect hands and arms. Soft leather gloves with canvas gauntlets to the elbow are the best for use.

(c) Boots:

- A pair of gum boots will protect the ankles and prevent bees from climbing up under trousers.

(d) Overalls:

- White overalls are occasionally worn. Light colored cotton materials arc preferable since they are cooler and create less risk for antagonizing bees

23. **Bee brush:**

- A soft-camel-hair brush is used to brush the bees off the honeycomb before it is taken for extraction.

24. **Smoker:**

- The smoker is used to calm bees and drive away bees from super.
- It consists of a metal fire pot with a funnel shaped cover and a bellow.
- A smoke releasing fuel (dried cow dung, hessian, waste jute bags or cardboard, old rag, wood shaving etc.) is burnt in the fire pot.
- Air is injected into the pot by operating the bellow and the smoke is directed to the desired spot.

25. **Decapping knife/ uncapping knife:**

- Single or double edged steel knife is used for removing wax capping from the honey comb before putting it in the honey extractor.

26. **Honey extractor:**

- It is invented by Frang von Hruschkain 1885.
- It consists of a cylindrical drum.
- A rack is fixed inside the drum to hold the supper frames.
- The rack is rotated by a set of gear wheels.
- The decapped honey frames are kept in the slots of the rack. The rack is rotated by operating the handle. Honey flow out from the combs by centrifugal force. The excreted honey comes out through the spout present at the bottom of the container.
- The honey comb is not damaged. So. it can be reused.

27. **Travelling screen/net:**

- It is a wooden frame with wire screen. It is highly useful for migration of honey bee colonies during hot summer season.

28. **Comb foundation mill:**
    - This is a machine to prepare comb foundation sheet used in beekeeping to make-bees build regular combs in frames that are convenient to handle.
    - J. Mehring of Germany made the first comb foundation in 1857.
    - Comb foundation is made by passing plain sheets of beeswax between two rollers that have the regular 3-faced cell base pattern embossed on them.
    - The patterns on the two rollers interlock properly, so that the 3-faced cell base on one roller matches with the base of each of the three cells on the other roller.
    - The distance between the rollers is fixed in such a way that a thin foundation is made that is readily accepted by the bees.
    - The rollers rotate on opposite sides.
    - The rotation is done by a handle attached to the lower roller. The cell size in the cell base pattern varies according to the size of the brood cells.

29. **Comb foundation sheet:**
    - It is a thin sheet of bee wax embossed with a pattern of hexagons of size equal to the base of the natural brood cells on both sides.
    - The size of the hexagon varies with bee species. For A. mellifera there arc 19 cells and for A. cerana 22- 23 cells/100 mm linear length.

30. **Embedder:**
    - It is a small tool with a spur or round wheel on the top. It is used to fix the comb foundation sheet on the wires of the frame.
    - Electric wire is also used for this purpose which is useful to reinforce the comb and give extra strength to the comb.

31. **Miscellaneous:**
    - Apart from these equipments, there are several miscellaneous equipment which are required from time to time viz., propolis screen, venom extractor, drip tray, swarm basket, wax melter, queen bee rearing equipment, comb foundation making equipment, honey straining, storage and processing equipment. etc.

5.11 DIVISION AND UNITING OF HONEY BEE BOXES

5.11.1 Division of honey bee boxes

Colony division is a method of multiplying bee colonies, i.e. producing two or more colonies from a mother colony. Colony division is used to control swarming as well as in commercial beekeeping to increase the number of colonies.

Methods for colony division

(i) Natural division using queen cells developed during swarming

The presence of multiple queen cells in a colony during the swarming season indicates a need for division. Dividing such colonies and using the queen cells in new daughter colonies can help control swarming. However, although it solves the immediate problem of swarming it does not help improve the genetic traits.

(ii) Colony division from queen production

Select the best colony based on the selection criteria given above. Produce queens from this colony before the onset of honey flow. These queens can be used to replace the old queen and to start new daughter colonies. The mother colony can be multiplied into several nucleus colonies but each should have at least 2 brood combs and 3–4 combs with food (nectar and pollen). The prepared colonies can then be sold or migrated according to need.

5.11.2 Uniting of Honey Bee Boxes

Uniting two colonies into one is done when one of them is weak or queen less or for other reason like bad traits etc. Each colony has its own colony specific odours and it is very difficult to combine the two colonies unless their odour is mixed well. Any attempt to unite these colonies without mixing their odour result in infighting and deaths will occur on large scale. Therefore, first step will involve bringing the two colonies into contact with each other. If uniting is done abruptly, the field workers of the colony shifted will not recognize the new place and returning to their original place will persist. This problem can be overcome by moving a hive gradually at the rate of two or three feet per day, so that the field bees get habituated to the changing position of their hive and will not drift back to the old site. When colonies are sufficiently close, one or two feet apart, they are ready for uniting. They can be united by three methods either (1) Direct uniting (2) Newspaper method (3) Smoking method.

(i) Direct uniting

The two hives to be united are brought near gradually and kept side by side. The queen with undesirable traits in one of the hives is removed. Next morning, when the bees are busy, the frames of two hives are gently put in one. The success of this method depends upon the skill with which it is done.

(ii) Newspaper method

Top cover is removed and the frames are covered with a piece of newspaper having a few holes made with a small nail the bottom, board of the upper colony is then removed and the brood chamber, is placed above the other colony, the newspaper forming a partition between the two. After a day or two, the odours of the colonies will mix and the bees will cut through the paper and will unite together, forming a single colony. After a few days all the frames can be placed in one hive and the upper chamber can be removed.

(iii) Smoking method

Colonies can be united using smoke method. When the colonies to be united have been brought close to each other, both should be smoked heavily and thin sugary syrup scented with oil of peppermint or wheat flour sprinkled over them. The combs with the bees of the colony to be united should be altered with the combs of the other colony. More smoke and syrup or flour should be applied and the colony closed. The work of the queen may be checked up after three or four days. It is better to unite a laying worker colony to several strong colonies by giving from one to two frames to each of them. If all its frames are united to one colony there is danger of latter's queen being killed by the laying workers.

5.12 SEASONAL MANAGEMENT OF HONEY BEE BOXES

Pollens and nectar are available only during certain period. When surplus food source is available is known as "honey flow season". In contrast during dearth period there will be scarcity of food. Suitable season for starting beekeeping coincides with mild climatic conditions and availability of bee flora in plenty. Normally, spring (February-April) and post-monsoon (Sep.- Nov.) seasons are the best periods to start beekeeping. Indian seasons are Spring season (Vasant ritu: mid February – mid April), Summer season (Grishma ritu: mid April to mid June), Mansoon season (Varsha ritu: mid June to mid August), Autumn season (Sharad ritu: mid August- mid October), Pre-winter season (Hemant ritu: mid October- mid December) and Winter season (Shishir ritu: mid December- mid February). Various operations required to be undertaken for augmenting colonies productivity are given below:

5.12.1 Spring Management

Management operations to be undertaken during spring are given below:

1. Examination of colonies

- ⊙ On some warm and sunny day, examine the colonies quickly and carefully with least exposure to the chilling weather and robber bees.
- ⊙ Unpack the colonies, clean the bottom board and replace the worn-out hive parts.
- ⊙ Assess the colony condition, working of the queen bee, brood rearing and food reserves.

- Provide early season stimulative sugar feeding (sugar: water =1:2), pollen or pollen substitute feeding to increase the foraging and brood rearing activity.

- Inspect the brood rearing. If there is no brood, the colony may be queenless. If there is less brood, the queen may be old and exhausted. Unite the weak/queenless colonies with the other colonies. If fresh dead bees are found, try to find out the cause. Bee mortality may be due to the bee disease or infestation of mite.

2. Equalizing the colonies

The colonies can be equalized by:

- Substituting the combs with food reserves/supplementary feeding.

- Providing the emerging bee combs.

- Uniting the bee combs/colonies.

- Giving young bees to the weaker colonies.

3. Provision of space

During spring, the colonies enhance the brood rearing. Hence, there should be no dearth of space to cope up with increased egg laying by the queen bee.

- Add good quality drawn combs (with worker cells) or frames with good quality comb foundations to the brood chamber as and when required.

- Avoid adding raised combs with too many drone cells.

- While providing super chamber, lure the bees to the super chamber with some bait in the form of a brood/honey comb.

4. Swarm prevention and control

Examine the colonies and remove congestion. Provide more drawn combs/comb foundations, supers, etc.

- Improve ventilation and provide required shade.

- Clip the wings of the laying queen.

- Use wire entrance guard/queen excluder at the bottom board.

- Reversing brood and honey chambers for mitigating congestion in brood chamber.

- Destruction of the queen cells raised due to swarming instinct,

- Dividing over-crowded colonies.

5. Control of mites and brood diseases

Examine the symptoms of various mites and brood diseases. On spotting any, take appropriate management measures to contain the menace.

6. Colony multiplication and commercial queen rearing

March-April is the best season for colony multiplication and commercial queen rearing. Improve the existing stock by selective breeding of best performing colonies. Mass reared queen bee can also be used for multiplication of existing stock and also for replacement of older queen bees (re-queening).

7. Extraction of spring season honey

Multiple extractions of honey during this period are possible. Only ripe honey from broodless combs from super chamber should be extracted.

5.12.2 Summer Management

Mid-April to June months is extremely hot. However, this is the major honey flow period too. The following operations need to be considered during the season.

1. Shifting the colonies to thick shade

Colonies should be moved to shady places every day by less than three feet.

2. Regulating the microclimate of the colonies

By using wet gunny bags over the colonies and sprinkling water around the colonies in the apiary during noon hours, temperature in the apiary can be reduced and humidity increased in hot and dry months of May and June.

5.13 PROVISION OF VENTILATION

Improve the ventilation of the colonies to cope up with the respiration of the bees and hastening the honey ripening by:

- Widening the entrance of the colony.
- Providing additional entrance in multi-chambered colonies.
- Staggering the chambers.
- Placing thin wooden splinters between two adjacent chambers for the circulation of fresh air.

4. Provision of fresh water

- Running water channel in the field.
- Cemented water reservoir tanks near the tube wells/pump sets with a sufficient number of sticks or wood pieces in the tank for the bees to sit on and lap the water.
- Earthen water bowls underneath the legs of hive stand also fulfill the water requirement of the colonies.
- An earthen pitcher with a small hole at its bottom is placed on a tripod and a slanting wooden plank is kept below the hole of the pitcher.

5. Honey extraction

Summer season honey can be extracted.

5.12.3 Monsoon Management

Manage the colonies during this season as below:

- Ensure colonies placement on upland area and away from village water ponds.
- Clean and bury deep the debris lying on the bottom board.
- Keep the surrounding of the colonies clean by cutting the unwanted vegetation which may hamper circulation of air.
- Provide sugar feeding (sugar : water=1:1), if required.
- Check robbing within the apiary.
- Unite weak/laying worker colonies. Control wax moth, ants, wasps and bee eating birds.

5.12.4 . Autumn Management

Important operations to be undertaken during this season, are:

- Provision of space.
- Strengthening the colonies to stimulate drone brood rearing, if queen bee rearing is to be undertaken.
- Control of ectoparasitic mites, brood diseases, wax moths and wasps.
- Autumn honey extraction before the winter sets in.

5.12.5 Winter Management

Normally winter extends from December to mid-February but this period may vary from region to region. During winter, very low temperature, westerly chilly winds, foggy/cloudy days and winter hamper the bee activity. Brassica comes in bloom during January. To perpetuate the colonies through winter, following operations are generally required:

1. Colony examination

Examine the colonies on a warm, sunny day for the presence of queen, brood and food reserves. Open the colony for minimum time to avoid chilling of brood. Weak colonies should be united with stronger ones so that the strong unit over-winters well.

2. Feeding

If there is food scarcity or expected in the ensuing winter, feed concentrated (sugar: water = 1:1) sugar syrup (supplementary feeding) by filling in the drawn combs at the onset of severe winter.

3. Shifting colonies to sunny places

The colonies should be shifted to sunny places with hive entrances facing south-eastwards.

4. Protection from the chilly winds

Plug cracks and crevices and narrow down the hive entrance.

5. Unite weak colonies with stronger ones

Follow newspaper method for uniting the colonies. 6. Removal of extra drawn combs and winter packing Remove the extra empty combs and store them properly to save them from mice/rats. Depending upon the strength of the colonies and severity of winter, provide one or two-sided inner winter packing combined with need based outer packing.

5.14 MIGRATORY BEEKEEPING (HONEY BEE BOX MIGRATION)

While preparing the honey bee colonies for migration, a number of points needed to be considered are given below:

5.13.1 Season

- Fasten the various hive parts and move the colonies during late evening, night or early morning, when all the bees are inside the hive, after closing the hive entrances with wire screen, ensuring required ventilation.

- In cold and rainy weather, the hives should be covered with a tarpaulin when being moved. Exposure to cold has the effect of causing bees to consume stores heavily to produce more heat and cluster together on the nest as they do in winter.

- During summer or monsoon season, colonies should be migrated during night when it is cooler and in hives with enough ventilation by exchanging the inner cover with traveling screen.

5.13.2 Distance of migration site

- Very long distance migrations of apiaries during winter at a stretch are possible provided the bees have sufficient food reserves and required ventilation. During cold weather, bees consume excessive food stores to produce heat and cluster together over the brood.

- The hive body and supers should be nailed or fixed properly to avoid their slipping enroute.

- During summer, it is better to have one or two halts/ journey breaks for short temporary sitting of the apiary for a day or two at some suitable place having some bee flora for easing out the confined bees. Moving the colonies continually for more than 48 h often leads to their brood mortality.

5.13.3 Number of Hives

If the number of colonies is small, it will not be economical to migrate them as the carriage/ transport charges per colony will be much higher than when the beekeeper has full vehicle load. However, in such cases, make it a full truck load by joining with other fellow beekeepers who also intend to take up migration to the same or nearby areas.

5.13. 4 Colony Strength

Bees are killed very often by overheating and lack of aeration but seldom by getting too cold. If weather is very hot and the colony is populous and hive is not spacious enough to allow expansion of the cluster, bees may very quickly smother/ get suffocated even when the top of the hive is covered with full wired travelling-screen. Thus, alternatively, the populous colonies may be divided and empty combs may be added for the expansion of the cluster in the hives before migration.

5.13. 5 Preparation and Packing of the Colonies for Migration

- ◉ Extract the surplus honey, if any, a few days prior to migration.
- ◉ All cracks and crevices in the hive should be sealed to bee-tightness.
- ◉ Excessively broken hive parts should be replaced with new ones.
- ◉ The hive body, bottom board and inner cover should be fastened together by stapling/ nailing.
- ◉ Always use two nails in slanting position on each side of every juncture. An alternative to the nails is to use metal or nylon travelling belts (migration belts) around the hive.

5.13.6 Type of the vehicle, and loading and unloading the colonies

- ◉ While migrating the colonies in vehicles such as trucks, the jerking movements will be forward and backward, hence, the length side of the hives should be kept parallel to the length of the vehicle.
- ◉ While loading the colonies in a tractor-trailer where the jerking movements are sideways, the bee hives (colonies) should be loaded with their length side parallel to the breadth of the vehicle or the axle of the vehicle.

5.13.7 Time of the day

If the whole of the apiary is to be shifted, it is better to move the bees in the evening or at night (when all the bees are inside the hive and temperature is low) or during rainy or cold weather when the bees are not foraging.

5.13.8 Timing in relation to flowering of crops

Colonies should not be taken to crop needing pollination until it is flowering sufficiently to be the predominant species in the locality. The delay in shifting colonies to the crop until flowering has begun, always increases pollination, particularly when the crop has short flowering period and is less attractive to bees than the other crops in the area. The same is true for honey production.

5.13.9 Placement of the migrated stock

- ◉ The migrated honey bee colonies should be sited away from the passages/ walkways where human or domestic animals' movements are expected.

- ◉ If migration is for pollination purpose, the bee colonies be placed within the crop and should be evenly distributed in the area to harvest the maximum pollination benefits and should not be crowded at one place.

6

ECONOMIC IMPORTANCE OF INSECTS

6.1 BENEFICIAL INSECTS

Insects are important because of their diversity, ecological role, and influence on agriculture, human health, and natural resources. Insects create the biological foundation for all terrestrial ecosystems. They cycle nutrients, pollinate plants, disperse seeds, maintain soil structure and fertility, control populations of other organisms, and provide a major food source for other taxa. Most major insect pests in agriculture are non-native species that have been introduced into a new ecosystem, usually without their natural biological control agents. Insects have evolved unique features in the animal world that are a surprise to experts in biomechanics and bioengineering because many are recent inventions of humans. Insects have been in competition with humans for the products of our labor ever since cultivation of soil began.

6.1.1 Pollinators of Crops (Bees, Wasps, Butterflies, Moths, Hoverflies, Beetles)

Many plants depend on insects to transfer pollen as they forage. Plants attract insects in various ways, by offering pollen or nectar meals and by guiding them to the flower using scent and visual cues. This has resulted in strong relationships between plants and insects. Value of crop production from pollination by native insects is estimated to be about $3 billion in US alone. When we talk about pollinators the ones that come to mind are honey bees and butterflies, but there are also many other insects that perform this job for flowering plants, as well. There are flies, wasps, beetles and even some other insects that most people know nothing about, such as Hemiptera and thrips. There are many important pollinating insect species in the orders: Hymenoptera (bees, wasps, and ants), Lepidoptera (butterflies and moths), Diptera (flies) and Coleoptera (beetles).

As adults these insects feed on pollen and nectar from flowers. They forage from plant to plant and may initiate pollination by transferring pollen from an anther to a stigma. Female bees and pollen wasps provision their nests with pollen and nectar that they actively collect onto their bodies. Their larvae then feed on the collected pollen and nectar. Yucca moth larvae do not feed on pollen or nectar but on the seeds of yucca plants. The adults pollinate the yucca plant by actively collecting pollen onto their palps and then placing the collected pollen on a receptive stigma to ensure proper seed set for their offspring.

Economic value of insect pollination worldwide is estimated at U.S. $217 billion (Science Daily, Sept. 15, 2008). German scientist found that the worldwide economic value of the pollination service provided by insect pollinators, bees mainly was dollar153 billion in 2005 for the main crops that feed the world. This figure amounted to 9.5% of the total value of the world agricultural food production. The study also determined that pollinator disappearance would translate into a consumer loss of food estimated between dollar 190 to 310 billion.

6.1.2 Predators of Pests (Dragonflies, Beetles, Bugs, Lacewings, Wasps)

The arthropods predator of insects and mites include beetles, true bugs, lacewings, flies, midges, spiders, wasps, and predatory mites. Insect predators can be found throughout plants, including the parts below ground, as well as in nearby shrubs and trees. Some predators are specialized in their choice of prey, others are generalists. Some are extremely useful natural enemies of insect pests. Unfortunately, some prey on other beneficial insects as well as pests. Major characteristics of arthropod predators:

- Adults and immature stages are often generalists rather than specialists.
- They generally are larger than their prey.
- They kill or consume many preys.
- Males, females, immature stages and adults may be predatory.
- They attack immature and adult prey.

Important insect predators include lady beetles, ground beetles, rove beetles, flower bugs and other predatory true bugs, lacewings and hover flies. Spiders and some families of mites are also predators of insects and mite pests. Natural enemies play an important role in limiting potential pest populations.

6.1.3 Parasites of Pests (Hymenoptera and Diptera)

Parasitoids are insects with an immature stage that develops on or in an insect host, and ultimately kills the host. Adults are typically free-living, and may be predators. They may also feed on other resources, such as honeydew, plant nectar or pollen. Because parasitoids must be adapted to the life cycle, physiology and defenses of their hosts, many are limited to one or a few closely related host species. Crop losses averted by beneficial insects from predators or parasites of agricultural pests are estimated to be $4.5 billion. The most valuable insect parasites belong to the following groups:

◉ Tachinid Flies (Diptera)

◉ Ichneumonid Wasps (Hymenoptera)

These parasites live in or on one host insect pest which is killed after the parasite completes its development. Parasite (also called parasitoid) adults are free-living; the immature stage lives on or inside a host and kills the host before the host completes its development. Parasites lay one or more eggs on the outside of the host body or they insert the eggs inside their host. The immature parasite feeds on the host and requires only a single individual prey to complete its development. Free-living adults may feed on nectar from flowering plants or obtain nutrients by piercing the body of host insects and withdrawing fluids (host-feeding). Parasites are often considered more effective natural enemies than predators because many have a narrower host range, require only one host to complete development, have an excellent ability to locate and kill their host and can respond rapidly to increases in host populations.

6.1.4 Productive Insects (Silkworm, Honey Bees, Lac Insects)

Sericulture is an agro-based industry. It involves rearing of silkworms for the production of raw silk, which is the yarn obtained out of cocoons spun by certain species of insects. The major activities of sericulture comprises of food-plant cultivation to feed the silkworms which spin silk cocoons and reeling the cocoons for unwinding the silk filament for value added benefits such as processing and weaving. Five varieties of silk worms are reared in India for producing this natural fibre. Bombyx mori, the silk worm, feeds on the leaves of mulbery to produce the best quality of fibre among the different varieties of silk produced in the country. Of the total production of 2,969 tonnes of silk in India, as much as 2,445 tonnes is produced by the mulberry silkworms, Bombyx mori.

Lac Insect any of the species of Metatachardia, Laccifer, Tachordiella,Austrotacharidia, Afrotachardina, and Tachardina of the superfamly Coccoidea, order Homoptera that are noted for resinous exudation from the bodies of females. Members of two of the families viz. Lacciferidae and Tachardinidae appear to be more concerned with lac secretion. There are several lac insects, some of which secrete highly pigmented wax. The Indian lac insect Laccifer lacca is important commercially. It is found in tropical or subtropical regions on banyan and other plants. The females are globular in form and live on twigs in cells of resin created by exudations of lac. Of the many species of lac insect,Laccifer lacca, (=Tachardia lacca) is the commercially cultured lac insect. It is mainly cultured in India and Bangladesh on the host plant, Zizyphus mauritianaand Z. jujuba. The insect starts its life as a larva or nymph which is about 0.6 mm long and 0.25 mm wide across the thorax. The young settles down on a suitable place of the host plant gregariously. On the average some 150 of such larvae may be present per square inch of the twig.

Apiculture or beekeeping is the maintenance of honey bee colonies, commonly in hives by a beekeeper in apiary in order to collect honey and beeswax, and for the purpose of pollinating crops. The genus Apis is comprised of a comparatively small number of species including the western honeybee Apis mellifera, the eastern honeybee Apis cerana, the giant bee Apis dorsata, and the small honeybee Apis florea.. Nectar is a sugar solution produced by flowers containing about 80% water and 20% sugars. Foraging bees store the nectar in the 'honey sac' where the enzyme invertase will change complex sugars into simple sugars called mono-saccharides. Upon return to the hive, the foraging bee will disgorge the partially converted nectar solution and offer it to other bees. Housekeeping bees will complete the enzymatic conversion, further removing water until the honey solution contains between 14 – 20% water.

6.2 INJURIOUS INSECTS

Less than 1% of insects are regarded as pests. They can be classified into the following categories.

6.2.1 Pests of Agriculture and Forestry (Locusts, Caterpillars, Bugs, Hoppers, Aphids Etc.)

Locusts are among the most destructive of all insect pests. Swarms of desert locusts were among the plagues of the Biblical Egyptians, and they still plague farmers throughout Asia and Africa. Their threat is so great that regional and international organizations monitor desert locust populations and launch control measures when necessary.Locusts are particularly destructive in hot, dry regions when a sudden increase in their numbers, combined with food shortage, forces them to migrate. They migrate in huge swarms, devouring virtually every green plant in their path.

6.2.2 Household Pests

(carpet beetles, furniture beetles, cloth moth, termites and silverfish) Common household pests include ants, termites, bed bugs, carpet beetles, furniture beetles, book lice, house flies, fleas, cockroaches, silver fish, clothes moths and spiders - the list seems almost endless. Common household pests enter our homes for shelter and food, and also to nest and breed. Common household pests can cause damage to our homes especially clothes, eatables and furniture. Household pests can also be a threat to health of our families by spreading bacteria, diseases or allergens in our homes. Household pests can be irritating, annoying or irritating and annoying. They can be controlled by spraying insecticides or by fumigants and by maintaining hygiene.

6.3 INSECTS OF MEDICAL AND VETERINARY IMPORTANCE (MOSQUITOS, FLEA, BEETLES, FLIES)

Mosquitoes can spread diseases such as malaria, yellow fever, dengue fever. Tsetse flies spread sleeping sickness. Lice suck human blood and can cause sores, which if left untreated can become infected which may lead to blood poisoning. Screw worm flies lay their eggs in the wounds of farm animals and pets. Horseflies and black flies suck blood and have painful bites, which can become infected. Houseflies spread germs and spoil meat by laying eggs in it. Bubonic Plague (or Black Death) was the worst disease epidemic in human history. It took 14 million lives–nearly 1 out of 4 people–in 14th- entury Europe. The plague is passed to humans by the bite of the Oriental rat flea Xenopsylla cheopis), which picks up the diseasecausing bacteria from rats.

Pests of stored grains: The most common insect pests of stored cereal grains are:

Rice Weevil (Sitophilus oryzae); Lesser Grain Borer (Rhyzopertha dominica); Rust Red Flour Beetle: (Tribolium spp.); Sawtooth Grain Beetle: Oryzaephilussurinamensis); Flat Grain Beetle: (Cryptolestes spp.); Indian Meal Moth (Plodiainterpunctella); Angoumois Grain Moth (Sitotroga cerealella); Khapra beetle(Trogoderma granarium); Rice moth (Corcyra cephalonica).

6.4 INSECT MANAGEMENT FOR STORED GRAIN DEPENDS UPON GOOD SANITATION AND GRAIN STORAGE PRACTICES

Clean storage areas to reduce the potential for insect migration into the new grain. Once the grain is dried to 13 percent moisture or less, cool it as soon as possible by running aeration fans. Reducing the grain temperature to less than 60°F stops insect reproduction, and lowering it to less than 50°F stops insect feeding activity. Infested grains should be fumigated by Aluminum phosphide (Phostoxin, Fumitoxin), which is best in most circumstances. Methyl bromide may also be used.

7

INSECT PESTS OF CEREAL CROPS AND THEIR MANAGEMENT

7.1 INTRODUCTION

The control of pests, diseases and weeds is one of the fundamental requirements for profitable cereal production. However, the large number of organisms that can threaten a crop makes management challenging. In addition legislative requirements in relation to chemical control measures are complex and ever changing. In practice, there is no substitute for knowledge and experience in identifying problems and choosing the most appropriate management technique for addressing them. The advice of a BASIS qualified agronomist can be invaluable in identifying threats to the crops and the most appropriate strategy for their control. In addition, a range of tools are available on the internet (or as hard copy books) to help identify weeds, pests and diseases and then to select the best plant protection product for the particular problem. The level of intervention a grower uses will very much depend on the particular circumstances of the farm and crop. At one end of the spectrum, certified organic growers will seek to avoid all chemical inputs; at the other end, some growers will routinely treat crops without paying much attention to the actual economic need for intervention. Most farmers lie somewhere between these two extremes.

Introduction of high yielding varieties, expansion in irrigation facilities and indiscriminate use of increased rates of agrochemicals such as fertilizers and pesticides in recent years with a view to increase productivity has resulted in heavy crop losses due to insect pests in certain crops. This situation has risen mainly due to elimination of natural enemies, resurgence of pests, development of insecticide resistance and out-break of secondary pests. Distribution, nature of damage, life history of important key pests of crops and their management strategies are outlined hereunder:

7.2 Insect Pests of Rice (Oryza Sativa)

7.2.1 Rice Stem Borer Scirpophaga Incertulas (Walker) (Pyraustidae: Lepidoptera)

Distribution: The yellow stem borer of rice attacks only rice and has wide distribution in all Asian countries.

Nature of damage: Larva feeds inside the stem causing drying of the central shoot called 'dead heart' in young plant or drying of the panicle called 'white ear' in older plants. October-December has been found conducive for the multiplication of the insect.

Life history: The female lays 15 - 18 eggs in a mass near the tip on the upper surface of tender leaf blade and covers them with buff coloured hairs and scales. A female lays about 2 - 3 egg masses and the incubation period ranges from 5 - 8 days. The newly hatched pale white larva enters the leaf sheath and feeds for 2 – 3 days and bores into the stem near the nodal region. Usually only one larva is found inside a stem but occasionally 2 - 4 larvae may also be noticed. The larva becomes full grown in 33 - 41 days and measures about 20 mm long. It is white or yellowish white with a well developed prothoracic shield. Before pupation it covers the exit hole with thin webbing and then forms a white silken cocoon in which it pupates. The pupa is dark brown and measures 12 mm long. The pupal period varies from 6 -10 days and may get prolonged depending on the weather conditions. The entire life-cycle is completed in 50 - 70 days.

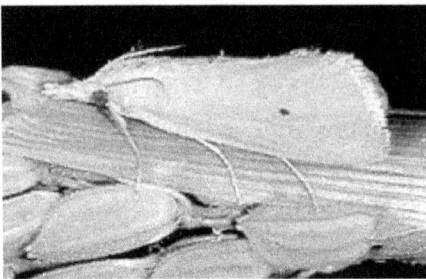

Rice stem Borer

Management Strategies:

- ⊙ Removal and destruction of rice stubbles from field and also collection and destruction of egg masses.

- ⊙ Clipping the tip of the seedlings prior to transplantation to eliminate egg masses.

- ⊙ Collection and destruction of moths using light traps.

- ⊙ Spraying of fenthion or fenitrothion or endosulfan or phosalone or monocrotophos or etofenprox or cartap hydrochloride or chlorpyriphos or phenthoate at 0.5 kg a.i./ha or fipronil 5% SC at 1 litre/ha if the economic threshold level of 10% dead heart is crossed in the nursery a week prior to pulling out the seedlings and the second after 15 days of transplantation.

⊙ An economic threshold level of 10% dead heart in vegetative stage and presence of 1 moth or 1 egg mass/sq.m. in the ear-head bearing stage has been suggested for adoption of chemical method of control by giving a third spray with one of the above chemical pesticides.

⊙ Seedlings root dip treatment for 12 or 14 hours before transplanting in 0.02% chlorpyriphos gives protection upto 30 days against stem borer.

7.2.2 Rice Gall Midge Orseolia Oryzae (Wood-Mason) Mani (Cecidomyiidae :Diptera)

Distribution: It is distributed throughout India. Five biotypes of gall midge have been observed. It is destructive in some parts of Kerala, Orissa, Andhra Pradesh, Madhya Pradesh and Bihar. It also breeds on a number of grasses such as Paspaladium geminatum, Paspalum scrobiculatum, Panicum spp., Ischaeum ciliare, Cynodon dactylon and Eleusine indica.

Nature of damage : The gall formed by this fly is popularly known as 'silver shoot' or 'onion shoot' or 'anaikomban' because of the formation of hollow pink or purple, dirty white or pale green cylindrical tubes bearing at their tips a green reduced leaf blade complete with ligules and auricles. It infests the rice even in the nursery but usually tillers are preferred. The loss in yield in a heavily infested crop may be up to 50 %.

Life history: The yellowish-brown fly, which is active at night, lays 100 - 300 reddish elongate tubular eggs singly or in groups of 2 - 6 on just below or above the ligules of leaf blade. The maggots hatch out in 3 or 4 days move down to shoot primordia in 6 – 12 hrs. The maggot feeds on the shoot primordia resulting in the suppression of the apical meristem and formation of radial ridges. Only one larva develops in a shoot apex and throughout its development it remains inside the tubular gall formed due to its feeding. Gall is the modified leaf sheath. The pale red coloured maggot feeds on growing point for 15 - 20 days and pupates inside the gall. The pupal stage lasts for 2 - 8 days and at the time of emergence of the adult the pupa wriggles upto the tip and projects half way out. The life cycle occupies 19 – 21 days but during winter it takes 32 – 39 days.

Rice Gall midge
(Source: www.agridept.gov.lk/)

Management Strategies:

- ◉ Seed treatment with chlorpyriphos 0.2% emulsion for 3 hours or seed mixing with either chlorpyriphos (0.75 kg a.i./100 kg seeds) or imidacloprid (0.5 kg a.i./100 kg seeds) provide protection for 30 days in the nursery.

- ◉ Seedling root dip in 0.02% chlorpyriphos emulsion before transplanting for 12 - 14 hours gives protection for 30 days.

- ◉ Removal and destruction of weeds that serve as alternate host plants.

7.2.3 Green rice leaf hoppers Nephotettix nigropictus (Stal.) and N. virescens (Distant) (Cicadellidae: Homoptera)

Distribution: These insects are found distributed in all rice growing areas in Asia and Africa. The insect is active during July – November in different regions.

Nature of damage: Both nymphs and adults suck the plant sap and cause browning of leaves. Both the species are known to be vectors of virus diseases of rice such as rice transitory yellowing and rice yellow dwarf .

Life history: The female of N. nigropictus is green and the male has two black spots extending upto the black distal portion on the fore wings. It has a black tinge along the anterior margin of pronotum and a submarginal black band on the crown of the head. In N. virescens black submarginal band on the crown is absent and the black spots on forewings do not extend upto the black distal portion. They also breed on some grasses. The female inserts the eggs in rows under the epidermis of leaf sheath and may lay upto 53 eggs. The life-cycle occupies about 25 days, the egg and nymphal periods being 6 – 7 and 18 days, respectively.

Management Strategies:

Spray application of phosalone or etofenprox or cartap hydrochloride or monocrotophos or acephate or chlorpyriphos or carbaryl, at 0.5 kg a.i./ha or fipronil at 50 g a.i./ha or application of granular insecticides such as phorate or sevidol or cartap hydrochloride or carbofuran at 1 kg a.i./ha or fipronil 0.3% G at 25 kg/ha.

7.2.4 Brown plant hopper Nilaparvata Lugens (Stal.) (Delphacidae: Homoptera)

Distribution: Distributed throughout South and South East Asia where rice is grown. It is known only to feed on rice and the weed Leersia hexandra Sw., Poaceae.

Nature of damage: It infests the rice crop at all stages of plant growth. Due to feeding by both nymphs and adults at the base of the tillers, plants turn yellow and dry up rapidly. At early infestation, round yellow patches appear which soon turn brownish due to the drying up of the plants and this condition is called 'hopperburn'. N. lugens is a phloem feeder. Very high infestation causes lodging of the crop resulting in yield loss ranging from 10 - 70 %.

Life history: Two forms viz., macropterous (long-winged) and brachypterous (shortwinged) are noticed and they are ochraceous brown dorsally and brown ventrally. The female inserts the eggs in two rows on either side of the midrib of the leaf sheath. The average member of eggs laid varies from 250 - 350. The incubation period is 6 - 9 days and the nymphal period is 10 – 18 days. The total life-cycle occupies 16 – 27 days.

Management Strategies:

- Spraying carbaryl (0.75 kg a.i./ha) or etofenprox, moncrotophos*, phosalone* or chlorpyrifos* @ 0.5 kg a.i./ha or lindane 20 EC at 1 litre/ha in the early stages of the crop.

- Application of granules of carbofuran at 0.75 kg a.i./ha or phorate at 1.25 kg a.i./ ha. * Application should be restricted to early stage of the crop.

7.2.5 Rice leaf folder Cnaphalocrocis medinalis Guen. (Pyraustidae: Lepidoptera)

Distribution: Occurs in all rice growing areas of our country. It is active from October to January. **Nature of damage:** The larva rolls the leaf blade by fastening its edges and sometimes even fastening the leaf tip to the basal part of the leaf blade and feeds from inside by scraping. In a severely infested field, the whole crop gives a sickly appearance with white patches. The infestation at boot leaf stage of the crop sometimes results in heavy loss of grain yield.

Life history: The brownish – orange coloured moth is small and has two and one distinct dark wavy lines on the brownish fore and hind wings respectively. Both wings have a dark brown to grey band on their outer margin. The eggs are laid singly or in pairs on the under surface of tender leaf blades. The incubation period is 4 - 7 days. The pale yellowish green larva becomes full grown in 15 - 27 days and pupates inside the leaf roll. Pupal period is 6 - 8 days. Total life-cycle varies from 25 - 42 days.

Management strategies:

- Removal of grass from field bunds.

- Need based spraying of phosalone or carbaryl or monocrotophos or etofenprox or cartap hydrochloride or quinalphos or fenthion at 0.5 kg a.i./ha or spray of fipronil 5 SC at 1 litre/ha.

7.2.6 Rice Earhead Bug Leptocorisa Acuta (Thunberg) (Alydidae=Coreidae): Hemiptera)

Distribution: This is one of the important pests of rice throughout India generally appearing before the flowering stage and continuing upto the milky stage. Apart from rice it also breeds on a variety of grasses.

Nature of damage: Both the nymphs and adults feed on the sap of peduncle, tender stem and milky grains making them turn chaffy. Life history: The female lays 250 - 300 eggs on leaf blade in long rows of 10 - 25 eggs and the incubation period is about a week. The slender greenish nymphs become adults in about two weeks. The longevity of the adults is 3 - 4 months.

Management Strategies:

(i) Dusting carbaryl 10 %, and repeat it depending upon the severity of infestation.

7.2.7 Whorl maggot Hydrellia philippina Ferino (Ephydridae: Diptera)

Distribution: It occurs in many rice growing countries in Asia. It occurs in severe form in certain high yielding varieties in Andhra Pradesh, Tamil Nadu and Orissa. It also breeds in Cynodon dactylon, Echinochloa crusgalli, E. colona, Fimbristylis miliacea, Eleusine indica and Paspalum scrobiculatum.

Nature of damage: The maggots attack the leaf blades even before unfurling and the initial damage is characterised by the presence of narrow stripes of whitish area in the blade margins. The tillers become stunted. Damaged leaves become distorted and may break off in the wind.

Life history: A female lays in 3 - 7 batches of about 100 cigar-shaped eggs singly on either side of leaves. Incubation period is 2 - 6 days. The larva undergoes 3 instars and the larval period is 8 - 17 days. Pupation take place in between the leaf sheath and the pupal period lasts for 5 - 9 days. There are 10 - 15 overlapping generations.

Management Strategies:

- Spraying of endosulfan or quinalphos or fenthion at 0.5 kg a.i./ha
- Alternatively, application of granules of carbofuran or fenthion at 0.75 kg a.i./ha.

7.3 PESTS OF WHEAT AND BARLEY

7.3.1 Brown Wheat Mite Petrobia lateens (Muller) (Arachnida: Tetranychidae)

Distribution: It occurs on wheat in the summer rainfall regions and expected to be severe where drought conditions are frequently encountered.

Nature of damage: The brown wheat mite feeds on sap from leaves by inserting two needle-like stylets into the leaf thereby withdrawing nutrients from the plant. During high mite populations the leaves may have a bronze appearance with some leaves even dying off as a result of intense feeding. They have a tendency to feed on the tips of the leaves, causing them to dry out and die. Heavily infested fields present a scorched withered appearance. Showers of 12mm or more may lead to the eradication of the mite population, but eggs present in the soil may start a new generation.

Life history: The mites are very small measuring about 0.5 mm in length, metallic brown to black with pale yellow legs and their forelegs are distinctively longer than the other three pair of legs. Eggs are generally laid beneath clods and are either active i.e. red in colour and not visible to the naked eye or dormant i.e. white eggs clearly visible on the underside of clods. Under favourable environmental conditions eggs hatch within 9 - 11 days. Dormant eggs may remain in the soil for long periods and hatch during July/August following light rainfalls. Dry conditions favour larval development and adulthood can be attained within 8 - 11 days. Only females occur and eggs are laid within 2 days after reaching adulthood. Subsequently, mite populations often reach pest status under dry conditions. The total life cycle is completed in 25.5 days.

Management strategies:

- ◉ Spraying formothion or oxydemeton methyl or phosphomidon @ 250 g a.i./ha. Spraying may be repeated after 15 days in case of severe infestation.

- ◉ Growing of tolerant wheat variety such as C-306.

7.3.2 Grain aphid Sitobion avenae (F.) (Aphididae : Homoptera)

Distribution: This pest is found distributed worldwide in parts of Europe, Asia, Africa, North America, Central America and Caribbean and South America. In India it is reported from Delhi, Haryana, Himachal Pradesh, Jammu and Kashmir, Madhya Pradesh, Punjab, Rajasthan, Uttar Pradesh and West Bengal. Nature of damage: A major pest of cereal crops in the spring. Heavy infestations can cause a reduction of the number of grains per ear and thus a noticeable reduction of the yield. Sooty moulds develop on the honeydew which they secrete. This aphid is also a vector of the barley yellow dwarf virus (BYDV). Life history: Primary hosts of this aphid belong to the genus Pubus. Secondary hosts mostly belong to the Poaceae (=Gramineae), including grasses like cocksfoot grass, and also cereals viz., oats, wheat, rye, barley. Winter eggs are laid on the straw of graminaceous plants and hatch in late winter, giving rise to several generations of apterous, virginoparous fundatrigeniae. Winged aphids emerge, spreading to other graminaceous plants and developing on the uppermost leaves before moving to ears as soon as these emerge. When populations are abundant or when the ripening grain hardens, winged aphids appear in a few days, leaving to create new colonies on Poaceae which are still green. During mild winters, this species survives parthenogenetically on new growth of winter cereals and other Poaceae.

Management strategies:

    i. Spraying formothion or oxydemeton methyl or phosphamidon @ 250ml a.i. ha.

    ii. Growing aphid resistant barley varieties DL117 or DL200.

    iii. Early sowing of barley by 1st fortnight of November usually reduces aphid attack.

7.3.3 Shootfly Atherigona naqvii (Muscidae: Diptera)

Distribution: Occurs in all wheat growing areas.

Nature of damage: The maggots bore into the shoot of young plants, a week after germination to about one month and as a result the central shoot dries up resulting in 'dead hearts'. If it is a little later the mother plant may produce side tillers. But the tillers also may be attacked. The infestation often goes as high as 60%.

Life history: The adult is a small dark fly. It deposits whitish eggs singly on the central surface of the leaves. The eggs hatch in 1 - 3 days and the maggots which are yellow in colour migrate to the dorsal surface of the leaf, enter the space between the leaf sheath and the axis and make a clean cut at the base of the leaf. The growing point of the plant dies and decays on which the maggots feed. The larval period lasts for 6 - 10 days. Pupation takes place inside the stem itself and the adults emerge in about a week. Each female fly is capable of laying 30 eggs during its life time. Life cycle occupies 17 - 20 days.

Management strategies:

- ⊙ Early sowing of wheat will reduce the attack.

- ⊙ In late sown crops (end of December to 1st week of January) spraying of cypermethrin 0.002% twice at fortnightly intervals in seedling stage was found effective. The most vulnerable period of crop is being upto 60 days of germination.

7.4 PEST OF MILLETS

Sorghum *(Sorghum vulgare)*

7.4.1 Sorghum shoot fly Atherigona soccata Rond. (Muscidae: Diptera)

Distribution: Occurs throughout the country.

Nature of damage: The maggots bore into the shoot of young plants, a week after germination to about one month and as a result the central shoot dries up resulting in 'dead hearts'. If it is a little later the mother plant may produce side tillers. But the tillers also may be attacked. The infestation often goes as high as 60%.

Life history: The adult is a small dark fly. It deposits whitish eggs singly on the central surface of the leaves. The eggs hatch in 1 - 3 days and the maggots which are yellow in colour migrate to the dorsal surface of the leaf, enter the space between the leaf sheath and the axis and make a clean cut at the base of the leaf. The growing point of the plant dies and decays on which the maggots feed. The larval period lasts for 6 - 10 days. Pupation takes place inside the stem itself and the adults emerge in about a week. Each female fly is capable of laying 30 eggs during its life time. Life cycle occupies 17 - 20 days.

Alternate hosts: The fly infests wheat, maize, small millets and grasses, besides sorghum.

Management strategies:

- ⊙ *A higher seed rate is adopted and the affected seedlings are pulled out and destroyed.

- Application of 10% phorate (Thimet) or carbofuran 3% granules at the time of sowing at the rate of 2.5 kg a.i./ha.

- Spraying of endosulfan @ 0.07% or cypermethrin @ 0.005% or cartap hydrochloride 0.5 kg a.i. /ha or triazophos @ 0.5 kg a.i. /ha twice a week after sowing or during second week.

* Seed rate of @ 12 kg/ha may be followed and the infected plants are removed.

7.4.2 Sorghum Stem Borer Chilo Partellus (Swinhoe) (Crambidae: Lepidoptera)

Distribution: It is found in all places of India where sorghum is grown. It is also found to attack finger-millet, maize, pearl-millet, sugarcane and wild grasses.

Nature of damage: Presence of 'dead heart' in the early stages is the main symptom. The bore holes may be visible in contrast to the dead-heart caused by the stem fly. Later it acts as an internode borer and is found till the time of harvest. Yield is affected much and the quality of the fodder is also reduced. The damage caused to the crop by this pest is estimated to range between 70 – 80 %.

Life history: The moth is medium sized and straw coloured. The female lays flat oval eggs, about 200 on the underside of the leaves, near the midrib. The eggs hatch in 2 - 6 days. The larva is pale white with black dots and brown head. It bores into the stem near the node and feeds upwards. The larvae remain dormant in winter and hibernate. The average number of caterpillars per plant is four. The larval period lasts from 28 – 50 days in summer to 193 days in winter. Pupation take place inside the stem and the adults emerge in 7 - 15 days depending upon the climatic conditions. Total life cycle is completed in 30 – 40 days.

Management strategies:

- Collection and destruction of the stubbles which are left in the field or heaped in one corner of the field since they act as a source of infestation, as the larvae hibernate in them.

- Spraying of carbaryl 0.1 % or endosulfan 0.07% thrice at an interval of 15 days from a month after sowing.

- Two whorl applications of 4 % endosulfan or 10 % carbaryl or 4% cartap hydrochloride granules, first at 5 kg /ha at 25 – 30 days after crop emergence and second at 10 kg/ha 10 - 15 days later. If infestation is severe, three applications at 5.0, 7.5 and 10.0 kg/ha are recommended.

7.4.3 Sorghum midge Contarinia sorghicola (Coq.) (Cecidomyiidae: Diptera)

Distribution: It has a world wide distribution and is considered to be one of the important pests of sorghum.

Nature of damage: The maggots feed on the developing grains and cause the developing grains to shrivel and severe infestation has a significant effect on the overall production of grains. The loss varies from 20 - 50 %.

Life history: The adult fly is very small, fragile and has a bright orange abdomen and a pair of transparent wings. The maggot feeds inside the developing grain and pupates there itself. It emerges between the tip of the glumes leaving the white pupal case attached to the tip in a characteristic manner. The life-cycle from egg to adult varies from 14 - 90 days.

Management strategies:

- ◉ Spraying of endosulfan 35 EC* 1 litre, or phosalone 35 EC 1 litre, or Malathion 50 EC 1 litre, or carbaryl 50WP 2 kg per hectare at nearly 90% ear-head emergence and repeated after 4 or 5 days.

- ◉ Phosalone 4% or endosulfan 4% or Malathion 5% or carbaryl 10% or quinalphos 1.5% dust at 12 kg/ha is also effective. * EC = Emulsifiable Concentrate.

7.5 PESTS OF SUGARCANE *(Saccharum officinarum)*

7.5.1 Early shoot borer *Chilo infuscatellus* Snellen (Crambidae: Lepidoptera)

Distribution: It is found to attack sugarcane in Tamil Nadu, Andhra Pradesh, Punjab, U.P., Bihar, West Bengal, Madhya Pradesh, Rajasthan and Maharashtra.

Nature of damage: The borer enters into young shoots and tunnels downwards. The upper portion of the central leaf whorl is thus cut off and dries up causing dead hearts in shoots from about a month old to 2 - 3 months crop. If the attack is in early stages the mother shoot dies completely and late attack induces profuse tillering.

Life history: The eggs are white and flat, laid in batches on the under surface of the leaves by the side of the midrib in three or more rows, one overlapping the other. Eggs are laid on the leaf sheath also. A moth may lay more than 200 eggs at a time and in each cluster 8 - 60 eggs will be found. The oviposition is at peak during May in March - April planted crop. The eggs hatch in 3 - 4 days. The caterpillars cut a hole on the side near ground level and enter the shoot and feed downwards. The affected tiller will decay emitting a foul smell. The caterpillar is white with five violet stripes on the dorsal side of the body with a brownish head. The crochets in the proleg are crescentic or semi circular. Each caterpillar migrates and attacks a number of shoots. The larval stage lasts for about 35 days and pupates inside the stem. The pupa is light brown in colour and the pupal period lasts for 10 days. The adult moth is small, pale grayish brown, the forewings with darker markings especially along the outer edge and the hind wings whitish. Total life cycle occupies 44 - 49 days.

Management strategies:

- ◉ Light earthing up of the tillers at the early stages of the crop (month old) during May and June reduces the incidence. A second earthing a month later reduces the borer attack considerably.

- Mulching with cane trash at the early stages also has been reported to reduce the incidence and help in conserving moisture.

- Cutting the affected tillers as close to the ground as possible and destroying them.

- Soil application of granules of cartap hydrochloride at 1 kg a.i./ha at planting followed by another application on 45th day for late planted crop.

- Cartap hydrochloride 4G, sevidol 8G and chlorpyriphos 10G at 1 kg a.i./ha as whorl application at 35th and 65th day are also found effective.

- Inundative releases of the egg parasitoid Trichogramma chilonis @ 50,000/ha from first month of planting at 7 - 10 days interval till one month prior to harvest.

- At 30th, 45th and 60th day of crop growth spray granulosis virus of Chilo infuscatellus (10-7 – 10-8 inclusion bodies/ml) at 500 l/ha.

- Release of 125 gravid females of Sturmiopsis inferens (Tachinidae) per ha at 45th day of crop growth.

7.5.2 Internode Borer Chilo Sacchariphagus Indicus (Kapur) (Crambidae: Lepidoptera)

Distribution: The insect is found throughout India and usually occurs on sugarcane late in its growing phase during June – December. It is serious in Andhra Pradesh, Karnataka, Kerala, Tamil Nadu and Uttar Pradesh. The larva has been found feeding on Pearlmillet, sorghum, maize, paddy and wild plants like Sorghum spontaneum, S. fusca and Pennisetum hoockenhackeri.

Nature of damage: It infests the crop after the third month when internodes begin to form and continues till the time of harvest. Symptoms will be visible only on close examination. The affected node will be thinner than the other nodes. Mostly the attack is found on the first five internodes. It feeds on the internal tissue making it turn red. The bore hole is plugged with excreta. Due to the attack the quality of juice is reduced and in severe cases cane formation is affected resulting in loss of weight.

Life history: Oval flattish eggs are laid in rows on the leaf. Generally, two rows, one just overlapping the other are laid. Each row will contain 9 - 11 scale like eggs. Egg period lasts about 3 days. The larva bores near nodes and feeds on the fibrous tissue by tunneling. It migrates and damages many nodes. The larva is white with dark spots on the body and four violet stripes on the dorsal surface. The dorsal median line is absent. The head is shiny yellowish brown. The crochets on the prolegs are circular. Larval period lasts about 30 days. Pupates in the leaf sheath and the pupal period lasts about a week. The adult is a pale brown moth with a white hind wing. Total life cycle lasts about 40 - 45 days.

Management strategies:

- Inundative release of the egg parasitoid Trichogramma chilonis @ 50,000 parasitoid/ha/week from the 4th up to 11th month after planting affords protection.

- Use of resistant varieties are CO 285, 453, 513, 617, 853, 915, 1007, 1287, 6806 and COJ 46.

7.5.3 Top Borer Scirpophaga Excerptalis Walk. (Pyraustinae: Lepidoptera)

Distribution: The insect has a wide distribution in India but is more serious in North India. It occurs on sugarcane in the later stages of growth during June – December and may persist till harvest. Wild Saccharum sp. is also attacked by this pest.

Nature of damage: Tunneling of midrib in a leaf, small holes in a parallel line in the freshly appeared leaves, dead heart which is reddish brown in the young crop (2 - 4 months old) and a bunchy top in a grown up crop are the characteristic symptoms. Sprouting of the eye buds affects the quality of the juice. Since the growth is restricted, the yield is also affected. It accounts for 20 - 30 % reduction in yield resulting in low juice quality and early maturity of the crop. The sucrose per cent in juice decreases by 2 units.

Life history: There are 5 broods reported in a year. The moth lays oval shaped flattish eggs in groups of 35 - 40, generally on the lower surface of the leaf and covers with crimson coloured hairs. The new larva hatches in 6 - 11 days and it bores through the mid rib and slowly reaches the central region. It feeds through the leaf whorl, enters the growing shoot from the top and feeds downwards. The caterpillar is cream yellowish white, very slender and soft. After entering the shoot it does not move out. Larval period lasts from 25 - 41 days. Pupates inside the burrow and emerges after 12 - 21 days. The adult comes out through the hole made by the larva before pupation. Moth is uniformly creamy white in colour. The first pair of wings of certain moths bears a black spot on each wing. Females have a crimson hairy tuft at the anal end. The total life cycle lasts 45 – 75 days. Heavy rains are usually conducive for the multiplication of the pest.

Management strategies:

- The egg masses and also the infested portions of plants may be collected and destroyed during the brood emergence period.
- Release of the ichneumonid parasitoid Gambroides javensis Rohw. has been found to be promising in Tamil Nadu.
- Variety Co 419 is comparatively resistant to this borer. Other resistant varieties are CoS 767, CoJ 67 and Co 1158.
- Soil application of carbofuran at 2 kg a.i. /ha or phorate at 1 kg a.i./ha for the third brood during first week of July is recommended.

7.5.4 Sugarcane Leaf Hopper Pyrilla Perpusilla Walker (Lophopidae: Homoptera)

Distribution: It is major pest of sugarcane in Punjab, Uttar Pradesh, Bihar and Maharashtra. Generally, it is a minor pest but sometimes assume a major form in different parts of India.

Nature of damage: Both the adults and nymphs are very active, jumping from leaf to leaf on slight disturbance. They suck the cell sap from the leaves and secrete honey dew which attracts the black fungus. Due to this feeding the leaves turns yellow and finally look withered and burnt with black encrustation. Life history: The female bug lays greenish yellow eggs in clusters, generally on the undersurface of leaves and between the detached leaf sheaths and the stem. The eggs are covered with white cottony, waxy filaments. 10 - 15 eggs are found in a cluster and they hatch in about a week. The nymphs have two characteristic anal processes and feed on plant sap. They develop into adults in 50 – 60 days. The insects are generally found from August - September. The intermittent periods of drought during July - September, heavy manuring and irrigation and luxuriant growth help in its profuse multiplication. It is also found that broad and soft-leaved varieties are more susceptible to this pest.

Management strategies: (i) Release of the lepidopteran ectoparasitoid Epiricania melanoleuca @ 4000 - 5000 cocoons or @ 4 - 6 lakh eggs/ha checks its multiplication. (ii) In case of severe infestation without the occurrence of the ectoparasitoid, spraying of endosulfan 0.07 % is quite effective.

7.5.5 Whitefly Aleurolobus Barodensis Mask. and Neomaskellia Bergii (Sign.) (Aleyrodidae: Homoptera)

Distribution: A. barodensis in recent years has assumed serious proportions on sugarcane in Bihar, Gujarat, Haryana, Karnataka, Maharashtra, Punjab, Tamil Nadu and Uttar Pradesh and in a mild form in Andhra Pradesh. It is serious on ratoon crops under water-logged as well as drought conditions.

Nature of damage: The sap drainage by nymphs results in the leaf gradually turning yellow and pinkish and ultimately the leaf dries up. The nymphs excrete large quantities of honey dew which accumulates on the affected leaves and the leaves appear black due to development of sooty mould interfering with photosynthesis. High infestation causes stunted crop growth and reduces juice quality. Severe whitefly infestation may result in reduction in cane yield up to 24 % and loss in sugar up to 2.9 units.

 Life history: The pale yellow adult female of A. barodensis lays about 80 eggs in linear masses on first and second top tender leaves. The eggs are fixed firmly on the leaf tissue on both the surfaces. The incubation period varies from 9 - 13 days. The nymphs feed by sucking the sap. The oval nymphs are black with fringes of wax and waxy deposit on body and the three nymphal instars last for respectively 4 – 7, 3 – 7 and 3 – 8 days. The fourth instar (pupal stage) lasts for 9 - 14 days. The last nymphal instar is the pupal stage from which through a T-shaped opening the adult emerges. The life-cycle is completed in 32 – 44 days.

Management strategies:

 i. Discourage ratooning in low lying areas and avoid water logging. Remove lower leaves containing pupae periodically.

 ii. Spray imidacloprid 0.01% or monocrotophos 0.05% or acetamiprid @ 0.01% after removing infested lower leaves. At least two or more sprays will be required at fortnightly intervals.

7.5.6 Sugarcane Woolly Aphid, Ceratovacuna Lanigera Zehntner (Pemphigidae: Homoptera)

Distribution: It was reported as a minor pest of sugarcane in Assam, Nagaland, Sikkim, Tripura, U.P. and Bengal in 1974. In July 2002 severe infestation of this pest in sugarcane crop was noticed in Maharashtra. Later this pest was found distributed in Karnataka, Andhra Pradesh, Tamil Nadu, Kerala, Uttrarakhand and Bihar. It is also reported to attack bamboo, Miscanthus sinensis.

Nature of damage: Nymphs and adults are found on the lower surface of the sugarcane leaves and suck the cell sap and excrete 'honey dew' which is dropped on the upper surface of the lower leaves. Honey dew encourages the growth of the fungus Capnodium spp. which results in black coating called 'sooty mould' on the upper surface of leaves affecting photosynthesis. Due to sap sucking, yellowish white spots develop on the leaves leading to drying of leaf edges and complete drying of leaves. Severe infestation causes mottling of leaves, stunted growth, and loss in sugarcane yield and sugar recovery. Losses up to 26% in sugarcane yield and 24% in sugar content have been reported.

Life history: Newly emerged nymphs are yellowish or greenish yellow in colour devoid of woolly filaments. Nymphs are found congregated at both the sides of the mid rib on the lower surface of the leaves. There are four instars observed after which they become adults. White coloured woolly filaments are observed on the dorsal side of the 3rd and 4th instar nymphs and not on the 1st and 2nd instars. Adults are black in colour and they possess two pairs of transparent wings and a pair of cornicles. Winged and wingless females are found to reproduce parthenogenetically throughout the year. Each female produced a maximum of 217 nymphs in 20 days. Nymphs completed four instars to become adults in 6 - 22 days. Overlapping generations are observed in the field.

Management strategies:

 i. Paired or wider row planting of sugarcane.

 ii. Release of the natural enemies Dipha aphidivora Meyr. or Micromus igorotus in shade net cages (5m x 5m) @ 50/cage and allow them to develop and disperse by removing the cages.

 iii. Release of D. aphidivora larvae @ 1000/ha and M. igorotus larvae @ 2500/ha.

 iv. Need based application of metasystox 0.0375% or endosulfan 0.05% or dimethoate 0.045% in case of severe infestation without the presence of natural enemies.

7.6 PESTS OF OILSEEDS

7.6.1 Groundnut (Arachis Hypogaea Linn.)

1. Red hairy caterpillars Amsacta albistriga Wlk. and Amsacta moorei Butler. (Arctiidae: Lepidoptera)

Distribution: This is a serious and devastating pest of rainfed groundnut crop. It is an endemic one and its seasonal outbreak in various areas is largely dependent on the climatic conditions and the local agricultural practices of the areas. Its outbreak takes place generally during May - June in Coimbatore district, during June - July in South Arcot, North Arcot and Salem districts and during August - September in Madurai and Ramanathapuram districts. It also occurs in Andhra Pradesh, Maharashtra and Karnataka.

Nature of damage: The larvae that hatch out from the eggs feed gregariously by scarping the under surface of tender leaflets leaving the upper epidermal layer intact. As they grow they feed voraciously on leaves leaving behind the petiole and midribs of leaves and the main stem of plants and may be seen marching from one field to another in thousands. Severely damaged crop presents the appearance as though the entire area has been grazed by cattle. Often it results in total loss of pods. Though it is principally a major pest of groundnut it prefers cowpea to a great extent. Apart from these crops it also feeds on sorghum, cotton, Pennisetum typhoideum L., Rhynchosia minima (Fabaceae), finger-millet, castor, etc.

Life history: The adults are medium sized moths. In A. albistriga the fore wings are white with brownish streaks all over and yellowish streak along the anterior margin and the hind wings white with black markings. A yellow band is seen on the head. In A. moorei the anterior marginal streak of fore wings and the band on head are red in colour. Both species are found together. After the receipt of heavy rains on the second evening at about 4 p.m. the moths emerge from their earthen cells in the soil. The moths mate and commence oviposition on the same day. The egg laying may last for 2 - 6 days. The creamy or bright yellow eggs are laid in groups mostly on the under surface of cowpea leaves usually sown along with groundnut as an inter crop and also on groundnut and occasionally on other vegetation, clods of earth, stones, dry twigs, etc. A female moth lays about 600 - 700 eggs but it has also been observed that as many as 2300 eggs have been laid by a moth. The incubation period ranges from 2 - 3 days. The larva becomes full grown in 40 - 50 days. It is about 5 cm long, reddish brown with hairs all over the body arising on warts. With the receipt of some showers, the grown up larvae burrow into the moist soil and pupate in earthen cells at a depth of 10 - 20 cm.

Management strategies: In view of the widespread out break of the pest, farmers need to adopt the control measures on a co-operative basis.

    i.   The pupae may be collected at the time of summer ploughings and destroyed.

    ii. Setting bonfires or light traps to attract the moths up to 11.00 P.M.

iii. Collection and destruction of egg masses should be carried out during the early stages of attack.

iv. A week after mass emergence of moths, the field should be dusted with phosalone 4% or carbaryl 10% dust to kill the first instar larvae which are vulnerable at this stage.

v. Grown up larvae are killed by spray application of phosalone 0.05 % or endosulfan 0.075%. (vi) Stray grown up larvae found in the field may be collected and destroyed. (vii)Nuclear polyhedrosis virus @ 250 LE/ha has been found promising in field scale control of the pest in Tamil Nadu.

2. Leaf miner Aproaerema odicella (Deventer) (Gelechiidae : Lepidoptera)

Distribution: The leaf miner is also one of the major pests of importance on groundnut crop all over India especially when raised under rainfed conditions. In the rainfed crop, the peak of attack is during September and October. It also infects Psoralea corylifolia, Cajanus cajan and soybean.

Nature of damage: Bunch variety is generally severely infested. The larva mines into tender leaflets or it webs together adjacent leaflets and feeds on the tissue. The leaflets get distorted and due to feeding get dried up in course of time. In a very severely infested crop, the whole field presents a burnt up appearance and the small adults could be seen flying in large numbers when one walks through the crop. The loss in yield of pods is also considerable.

Life history: The moth is very small with dark brown wings and small distinct white spot on fore wings. It lays shining, sculptured eggs, singly on tender leaves. The incubation period is 3 days. A moth on an average lays up to 200 eggs. The larvae immediately after hatching mine into the leaves. Later they come out and web together the leaves and feed on the green matter. Larval period lasts for 9 – 17 days and the larvae pupate in the leaf folds. Pupal period is 3 – 7 days.

Management strategies:

i. Dusting phosalone 4% or carbaryl 10% or spraying fenitrothion 0.025 % or phosalone 0.05% or monocrotophos 0.05 % or chlorpyriphos 0.05 %.

7.6.2 Mustard (Brassica Campestris)

1. Mustard sawfly Athalia lugens proxima (Klang) (Tenthredinidae: Hymenoptera)

Distribution: This is one of the very few hymenopterous insects noticed to infest Cruciferous crops all over India. It is a pest of cold weather, generally active during October to March.

Nature of damage: The larvae feed voraciously on leaves. Apart from mustard it also attacks radish and allied plants. It feeds during mornings and evenings from the margin of the leaf towards the centre. During day time it prefers to stay in the soil.

Life history: The adult is black with yellowish femora and thorax. The female possesses a saw-like ovipositor and inserts the eggs very near the leaf margin. A female on an average lays about 60 eggs. The larva is cylindrical and dark gray with three pairs of thoracic legs and seven pairs of prolegs on abdominal segments 2 -8. Its body surface is hairless. Young larva is greenish grey in colour and its colour becomes darker in the later instar. It measures about 15 – 20 mm long and pupates in an earthen cocoon in the soil. The egg, larval and pupal periods occupy 4 – 5, 13 – 18 and 10 – 15 days, respectively. Parthenogenetic development is also observed.

Management strategies:

    i.   Spray application of carbaryl 0.1 % or endosulfan 0.07 % or phosalone 0.05% or profenofos 0.05%.

2. MUSTARD APHID Lipaphis erysimi Kaltenbach (Aphididae: Homoptera)

Distribution: The most serious pest of the mustard crop in India. Besides brassicas to which mustard belongs, this pest attacks a number of other economic plants, particularly those of the family Cruciferae. Like many other important aphid pests, this species has a very wide distribution in the world.

Nature of damage: These aphids are small (about 2 mm), generally globular with piercing and sucking mouth-parts. They possess a pair of small tubular structures at the posterior region of their body, called cornicles. It pierces its proboscis into the tender plant tissue and sucks the plant sap. It excretes honeydew that covers practically the whole surface of leaves and the tender shoots. A black mould develops on the honeydew which interferes with the photosynthetic activities of the plant.

Life history: The aphid population generally makes its appearance sometimes during winter and it continues to breed parthenogenetically till the end of spring when winged individuals are produced and large-scale dispersal takes place. The population, however, dwindles mostly due to climatic reasons and practically disappears for the whole of the summer and also most of the autumn.

Management strategies:

    i.   Early sowing of mustard before 15th October will help to escape the attack of the pest and economic damage. (

    ii.   Spray application of metasystox 0.05% or imidacloprid 0.01% or acetamiprid @ 0.01%.

7.6.3 Coconut (Cocos Nucifera)

1. Rhinoceros beetle Oryctes rhinoceros L. (Scarabaeidae: Coleoptera)

Distribution: It is widely distributed in India and persistent in all coconut growing areas. Pineapple, sugarcane, Aloe, African oil palm, palmyrah, date palm and talipot palm are also attacked by the beetle.

Nature of damage: The damage is inflicted by the adult beetle which burrows by remaining in between leaf sheaths near the crown and thus cut across the leaf in its folded condition. The damaged leaves show characteristic clipping or holes in the leaflets. Frequent infestation results in stunting of trees and death of growing point in young plantations. The infestation can be easily made out by the chewed fibrous material present near holes.

Life history: The adult beetle is stout, black or reddish black, about 5 cm long and has a long horn projecting dorsally from the head in male; in the female the horn is short. The female lays the oval, creamy white eggs in manure pits, decaying vegetable matter, undisturbed heaps of cattle excreta etc. to a depth of 5 - 15 cm. A female may lay up to 140 eggs and the incubation period is 8 – 18 days. The grub feeds on the decaying matter and the larval stage lasts for 99 - 182 days. It is stout, sluggish and white in colour with a pale brown head and is usually found at a depth of 5 - 30 cm. The grubs pupate in earthen cells at a depth of 0.3 - 1 meter and emerge as adults in 10 – 25 days. The total life-cycle from egg to adult ranges from 3.5 - 8 months and the adult longevity extends up to 290 days.

Management strategies:

i. The grubs in their breeding places should be killed by spray application of carbaryl 0.1 % solution at least once in three months.

ii. The beetles should be extracted from the crown with the help of iron hooks and a mixture of sand plus carbaryl 10% dust in equal proportions should be filled in the axils of innermost 2 – 3 leaves on the crown twice a year during pre and post monsoon periods.

iii. The grubs are susceptible to the fungi Metarrhizium anisopliae and Beauveria bassiana and so application of these in breeding sites is recommended.

iv. Pieces of tender coconut stem split longitudinally and treated with fresh toddy or 1 kg castor cake soaked in water in small mud pots when kept in coconut gardens are found to attract the beetles. And these could be used in poison baits.

2. Red Palm evil Rhynchophorus ferrugineus (Curculionidae : Coleoptera)

Distribution: This is one of the important pests of coconut in India in Kerala, Karnataka, Goa, Tamil Nadu and Andhra Pradesh and its attack often results in the death of the palm. In addition to coconut other palms are also attacked by the weevil.

Nature of damage: Few small holes with protruding chewed fibrous material and oozing out of a brown liquid from such holes, present in the tree trunk, indicate the early infestation by the pest. In the advanced stage of attack the central shoot shows sign of wilting and a large mass of grubs, pupae and adults of the insect could be seen inside the trunk at the affected portion. In the grown up trees the crown region alone is infested.

Life history: The reddish brown weevil has six dark spots on thorax and in the male the conspicuous long snout has a tuft of hairs. The female lays the eggs in scooped out small cavities on palms of up to 7 years and on older trees they may be deposited in the wounds and other cut injuries found on the trunk crown. A female lays as many as 276 eggs which are oval and white. The incubation period is 2 - 5 days. The apodous light yellowish grub with a red head becomes full grown in 36 - 78 days and pupates in a fibrous cocoon inside the trunk itself. It emerges as an adult in 12 - 33 days.

Management strategies:

i. The dying and already damaged palms should be destroyed and as far as possible inflicting mechanical injuries on trees should be avoided.

ii. The infested portion should be scooped out and dressed with tar. A solution of 1 % pyrocone E (a mixture of pyrethrin 1 part + piperonyl butoxide 10 parts) i.e. 1 part in 100 parts of water, or 1 % carbaryl or monocrotophos 36 WSC 5 ml + DDVP 76 WSC 5ml when injected through a hole on the crown at 1000 - 1500 ml per grown up tree brings about appreciable control of the pest.

3. Black-headed caterpillar Opisina arenosella (Wlk.) (Cryptophasidae: Lepidoptera)

Distribution: One of the pests of importance on coconut all over peninsular India but more injurious along the east and west coasts.

Nature of damage: The larvae live on the under surface of leaflets within galleries of silk and frassy material and feed by scraping the green matter. In case of severe attack due to large scale drying of leaflets the whole plantation presents a burnt up appearance from a distance.

Life history: The grayish white small moth lays about 130 eggs in groups on leaves and the larvae hatch out from the eggs in about 5 days. The larval period lasts about 40 days and the caterpillar is greenish brown with dark brown head and prothorax, and a reddish mesothorax. It pupates inside the web itself in a thin silken cocoon and after about 12 days emerges as adult.

Management strategies:

⊙ The infested fronds should be cut and burnt.

⊙ In the case of young trees carbaryl 0.1% may be sprayed.

⊙ Trunk injection of monocrotophos 36 WSC at 5 ml/palm is also effective.

⊙ Root feeding with monocrotophos is suggested. For this select a mature dark brown root, cut with a sharp knife and immerse the cut end in an emulsion containing 20 ml of 1:1 monocrotophos 36 WSC and water in a well secured small polythene bag (15 cm x 10 cm). To avoid residues in the tender coconuts, this treatment is not suggested for fruiting trees.

⊙ Periodically releasing of its parasitoids such as Goniozus nephantidis, Bracon brevicornis, Elasmus nephantidis and Trichospilus pupivora is recommended.

7.6.4 Sunflower (Helianthus Annuus L.)

1. Sucking pest, Amrasca biguttula biguttula Ishida (Cicadellidae: Homoptera)

Distribution: This pest is of economic importance in Maharashtra, Tamil Nadu and Karnataka causing crop loss up to 46 %. Though it may appear on the crop round the year, it is serious during certain months at different places. Summer crops are likely to suffer more with this pest than kharif crop.

Nature of damage: The incidence would start from seedling stage and prevail right through entire plant life. Stunted growth of plant, cupped and crinkled leaves, burnt appearance of leaf margins are symptoms of damage.

Life history: The female lays on an average 15 eggs into the spongy parenehymatous tissue between the vascular bundles and the epidermis and they hatch in 4 - 11 days. The nymphs moult five times and the whole life cycle is completed in two weeks to more than a month and a half depending upon the temperature and humidity prevailing in the field.

Management strategies:

i.   Insecticides like phosphamidon (0.03%) or dimethoate (0.03%) or monocro-tophos (0.05%) or imidacloprid (0.01%) may be sprayed @ 650-700 litre spray solution per hectare if the pest build up is very high.

ii. Seed treatment with imidacloprid @ 5g and 7.5 g/kg of sunflower seed protects from jassid up to 35 - 40 days after sowing.

2. Capitulum borer, Helicoverpa armigera (Hubner) (Noctuidae : Lepidoptera)

Distribution : The capitulum borer, H. armigera is highly polyphagous with about 181 host plants including important crop plants such as pulses, cotton, vegetables, oilseeds etc. and the pest is prevalent throughout India.

Nature of damage: The larva is capable of developing on foliage which is rather less common in field's situations. On a bloom, usually, larvae on hatching would get into the bottom of the preripheral florets and feed on ovaries. During pre-anthesis stage they feed scraping the bracts first and later feed through ray-florets which cover disc florets and finally find access to immature ovaries. The larval growth is better supported by developing seeds.

Life history: H. armigera passes through four generations in Punjab and seven to eight generations in Andhra Pradesh. Several crops like maize, sorghum, cotton, sunflower, tomato, pigeonpea, chickpea etc., are found to support large populations of H. armigera.

Emergence of H. armigera moth has been observed evening any time after 1600 hrs. the peak emergence being between 20.00 and 22.00 hr. Pre-oviposition period ranged from 1 - 4 days, oviposition period 2 - 5 days, and post oviposition period 1 - 2 days. Each female moth can lay on an average 700 - 1000 eggs. The incubation period ranges from 2 - 5 days. There are normally six instars, but exceptionally seven instars are found in cold season. The larval period ranges from 8 to 33.6 days with 8 to 12 days on tomato,

21 - 28 days on chickpea, 21 - 28 days on maize, 33.6 days on sunflower and 20 - 21 days on cotton. The full grown larvae pupate in earthen cocoons in the soil. Pupal period vary from 5 - 8 days in India.

Management strategies:

i.   A significant reduction in pest density is achieved with the spray of NPV @250 LE*/ha.

ii. NSKE (5%) and many neem origin pesticides are found effective in reducing damage due to H. armigera.

iii. Endosulfan (0.05%) on 25 and 45 DAS is ideal for management of this pest in a short duration variety like Morden.

iv. Endosulfan (0.05%), cypermethrin (0.005%), fenvalerate (0.005%) and deltamethrin (0.002%) spray @ 650 – 700 litre/ha against the head borer are found to be effective.

v.   * LC = Larval Equivalent

3. Tobacco caterpillar Spodoptera litura (Fabricius) (Noctuidae: Lepidoptera)

Distribution: It is cosmopolitan, highly polyphagous and is reported on sunflower in all sunflower growing areas.

Nature of damage : Early instar larvae (gregarious phase) scrape on green matter that give a mesh like appearance to damaged leaves which can be spotted easily from a distance. Older larvae cause total defoliation.

Life history: Adult moth has dull brown forewings with white markings, hind wings are hyaline. Eggs are laid underneath the leaves in clusters (200 - 300 eggs) covered with cream coloured hairs and scales. Egg period is 3 - 4 days. Larvae are gregarious when young, later disperse having 5-6 instars. Larval duration ranges from 15 - 28 days. They feed on foliage at night, hide in soil and debris during day. The larvae pupate in the soil in an earthen cell. Pupal period is 7 - 10 days.

Management strategies:

◉ Monitoring of moth activity through pheromone traps.

◉ Collection and destruction of egg masses and gregarious early instars present on undersurface of leaves.

◉ Spray of monocrotophos 0.05% or dichlorvos 0.05% or cypermethrin 0.005% in 500 litre water/ha in case of severe incidence.

◉ Use of poisoned bran bait (125 ml monocrotophos + 1 kg jaggery + 10 kg rice bran) is effective against later instars.

4. Bihar hairy caterpillar Spilosoma oblique Walker (Arctiidae: Lepidoptera)

Distribution: It is highly polyphagous and mainly a pest of rabi-summer sunflower in Maharashtra.

Nature of damage: The larvae are foliage feeders. Early instars feed on chlorophyll and later instars defoliate the crop. Drying up of infested leaves is a characteristic symptom.

Life history: Adult moth lays eggs in clusters. Larvae are hairy, gregarious in early instars and disperse later. Larval period varies from 14 - 21 days. Pupal diapause is noticed. Generation time is 38 - 164 days Management strategies: (i) Collection of infested leaves which show characteristic drying symptoms will reduce the population to a great extent because of the gregarious nature of young larvae. (ii) Spraying contact insecticides endosulfan or quinalphos or carbaryl at 0.05 – 0.1%.

5. Green semiloopers Trichoplusia ni, Thysanoplusia orichalcea (Fabr.) (Plusia orichalcea) (Noctuidae: Lepidoptera)

Distribution: Regular pest of sunflower in Maharashtra, Karnataka during August and September. It is also found to attack cotton, legumes, solanaceus plants, sweet potato and some cucurbits.

Nature of damage : Early instars feed on chlorophyll of tender leaves causing transparent leaf spots, later feed from leaf margin and defoliate leaving midribs in case of severe incidence.

Life history: Larvae are green in colour with a thin white lateral line and two white lines on the back, active and form loop in motion; swollen at posterior end and tapers anteriorly. Pupate in white transparent silken cocoons in leaf litter or crop debris. Life cycle takes 30 days.

Management strategies:

    i.   Spray quinalphos 0.05% in case of severe incidence.

7.7 PESTS OF FIBRE CROPS

7.7.1 Cotton (Gossypium Hirsutum, G. Herbaceum, G. Barbadense And G. Arboreum)

1. Cotton aphid Aphis gossypii Glover (Aphididae: Homoptera)

Distribution: It has a world wide distribution. It also attacks Lady's finger, brinjal, guava, gingelly etc.

Nature of damage: The greenish brown soft bodied small aphids infest the tender shoots and the under surface of leaves in very large numbers and suck the sap. Severe infestation results in curling of leaves, stunted growth and gradual drying and death of young plants. Black sooty mould develops on the honey dew of the aphids which falls on the lower leaves affecting photosynthetic activity. The Economic Threshold Level (ETL) is 10% affected plants counted randomly.

Life history: The alate as well as apterous females multiply parthenogenetically and viviparously. In a day a female may give birth to 8 – 22 nymphs which become adults in 7 – 9 days.

Management strategies:

i. Spray application of dimethoate 0.03 % or methyl demeton 0.025 % or mono-crotophos 0.04 % or imidacloprid 0.01 % affords protection.

ii. Include also imidacloprid seed treatment for sucking pests @ 3 – 5g /kg seed that protects the crop around 30 – 45 days or so.

2. Cotton leaf hopper Amrasca biguttula biguttula Ishida (Cicadellidae: Homoptera

Distribution: It is distributed in all cotton growing areas. It is also found to breed on brinjal, potato, lady's finger and sunflower.

Nature of damage: Both the nymphs and adults suck up the plant sap from the under surface of leaves. The leaves show symptoms of "hopper burn" such as yellowing, curling, bronzing and sometimes drying up, and these symptoms are expressed differently depending on how the different varieties react to the toxic saliva of the insect. The vigour of the plants is impaired to a great extent. The ETL is 2 Jassids or nymphs per leaf or yellowing in the margins of the leaves.

Life history: The female leaf hopper inserts about 15 eggs inside leaf veins and the incubation period ranges from 4 to 11 days. The nymphal period occupies 7 - 21 days depending on the weather conditions.

Management strategies:

i. Spray application of dimethoate 0.03 % or methyl demeton 0.025 % or mono-crotophos 0.04 % or imidacloprid 0.01 % affords protection.

ii. Include also imidacloprid seed treatment for sucking pests @ 3 – 5g /kg seed that protects the crop around 30 – 45 days or so.

3. Cotton whitefly Bemisia tabaci (Gennadius) (Aleyrodidae: Homoptera)

Distribution: It is widely distributed in India and particularly serious on cotton and brinjal. It is very important as a vector of leaf curl virus disease of crops like tobacco, cotton, etc. and vein clearing disease of lady's finger. It also breeds on a variety of plants such as hollyhock, lady's finger, tobacco, safflower, Achyranthes aspera, Lab – lab niger, topioca.

Nature of damage: In cotton the nymphs are found in large numbers on the under surface of leaves and drain of sap due to sucking. Severe infestation results in premature defoliation, development of sooty mould on honey dew excreted, and shedding of buds and bolls and bad boll opening. The ETL is 5- 10 nymphs or adults per leaf before 9 A.M.

Life history: The female whitefly lays the eggs on the under surface of tender leaves. The egg and nymphal periods occupy respectively 3 - 5 and 9 - 14 days during summer and 5 - 33 and 17 - 73 days in winter. The pupal period is 2 - 8 days. The total life-cycle ranges from 14 - 107 days depending upon the weather conditions.

Management strategies:

    i.   Spray application of acephate 0.075 % or imidacloprid 0.01 % or acetamiprid 0.01% or neem oil 0.3 % brings about control of the pest.

4. Spotted bollworms Earias insulana Boisd. and E. vittella (Noctuidae : Lepidoptera)

Distribution: Widely distributed in India, Myanmar. It is also found to attack Bhindi.

Nature of damage: The initial infestation takes place on 6 week old crop in which the larva causes detopping (drooping and drying of the shoot) due to its feeding by boring into it. In the later stages of the crop, the buds, flowers and bolls are damaged and a larva may migrate and attack fresh parts. Heavy shedding of early formed flower buds due to the pest is a common feature in cotton fields. The lint from attacked bolls will not be clean. The ETL for this pest is 5% damaged fruiting bodies or 1 larva per plant or total 3 damaged squares / plant taken from 20 randomly selected plants.

Life history: The moth of E. vittella has green fore wings with a white streak on each of them whereas that of E. insulana is completely green. The female moth deposits 2 or 3 eggs on bracts, leaf axils and veins on the under surface of leaf. The egg is crown-shaped, sculptured and deep sky blue in colour. A female may lay about 385 eggs and the incubation period is about 3 days. The larva becomes full grown in 10 – 12 days. The larva of E. vittella is brownish with a longitudinal white stripe on the dorsal side and without finger-shaped processes on its cream coloured body and orange dots on prothroax. The boat shaped tough silken cocoon is dirty white brownish and may be found on plants or on fallen buds and bolls. The pupal period is 7 – 10 days. The total life cycle ranges from 20 to 22 days.

Management strategies:

    i.   The infested portions as well as shed buds and bolls should be removed and destroyed. (ii) Periodical spray application of compounds like phosalone 35 EC @ 1.5 to 2.5 l/ha or carbaryl 50 WP @ 2.5 to 3.0 kg./ha or endosulfan 35 EC @ 1.5 to 2.0 l/ha or monocrotophos 40 SC @ 1.0 to 1.25 l/ha or profenofos 50 EC@ 0.75 to 1.0 kg /ha or thiodicarb 75 WP@ 625 g/ha etc. has been reported to be effective.

    ii. The synthetic pyrethroids fenvalerate and permethrin @ 100 - 150 g a.i./ha, cypermethrin @ 80 g a.i./ha, and deltamethrin @ 12.5 to 15 g a.i./ha are very effective in controlling the bollworms of cotton which may be alternated with other groups of insecticides.

5. Pink bollworm Pectinophora gossypiella Saund. (Gelechiidae: Lepidoptera):

Distribution: This is a well known pest of cotton found distributed all over the world. Alternate host plants of this pest are Lady's finger, hollyhock and Thespesia populnea, etc.

Nature of damage: The larva enters the developing boll through the tip portion and the entrance hole gets closed up as the boll matures. It feeds on the seeds and moves to adjacent locule by making a hole through the septum. The infested flower buds shed prematurely. A typical rosette-shaped bloom when examined will contain the larva. The infestation results in the seeds being destroyed in addition to retardation of lint development and weakened lint and staining of the lint both inside the boll and in the gin. Further, infested bolls open prematurely and expose it to invasion by saprophytic fungi. The seeds from damaged bolls show lower germination. The infestation ranges from 40 to 85 %. The ETL for pink bollworm is 8 months / trap per day for 3 consecutive days and the traps are to be installed @ 5/ha. or 10% infested flowers or bolls with live larvae.

Life history: The adult is a small dark brown moth and a female lays flattened and striated eggs on the bolls or in between bracts or on buds and flowers, the average being 125 eggs. The egg period varies from 4 - 25 days. The 15 mm long pinkish larva with dark brown head and prothoracic shield becomes full grown in 25 - 35 days and pupates in a thin silken cocoon among the lint, inside a seed or in double seeds, in between bracts or in cracks in the soil. The pupal period is about 6 – 20 days. Both short – cycle larvae and long-cycle larvae occur in Northern India and hibernation during winter takes place in the larval stage. In South India the insect is not known to hibernate in any stage of its development.

Management strategies:

i.   Need based spray application of carbaryl 50 WP @ 2.5 kg/ha or quinalphos 25 EC @ 2 - 3 l/ha or profenofos 50 EC @ 1.5 - 2 l/ha based on ETL.

ii.   Fumigation of seeds with methyl bromide at 1.5 kg/100 cu. m. for 24 hours or with aluminum phosphide at 18 tablets/100 cu. m. or heat treatment for a few minutes at 60°C kills hibernating larvae in seeds.

7.7.2 Jute (Corchorus Olitorius & C. Capsularis)

1. Stem girdling beetle Nupserha bicolor postbrunnea Dutt (Lamiidae: Coleoptera)

Distribution: Corchorus olitorius and C. capsularis are the two species of Jute cultivated in Assam, West Bengal, Orissa, Tripura and Uttar Pradesh which are attacked by this pest. It is particularly serious on C. olitorius.

Nature of damage: The adult beetle girdles the stem at two levels before it starts oviposition. This causes withering, drooping and death of the portion above the lower girdle to a length varying from 5 - 50 cm thus resulting in loss of fibre yield.

Management strategies:

i.   Spray application of phosalone 0.07% or endosulfan 0.07 % at fortnightly interval.

ii.   Removal and destruction of drooping stem portions and stem casings containing the larvae in diapause.

iii. Growing of C. capsularis jute which is not preferred by the insect.

2. Jute weevil Apion corchori Mshll. (Apionidae: Coleoptera)

Distribution: The insect occurs in a serious form on jute in Bihar and U.P.

Nature of damage: The adult weevil excavates a small hole on the stem and oviposits. The grubs tunnel into the pith. Due to damage a gall-like swelling is formed. C. capsularis is more susceptible to attack than C. olitorius.

Life history: The weevil is dull black or dark brown in colour and a female is capable of laying up to 675 eggs during an oviposition period of about 4 months. The egg, larval and pupal periods last for 3 – 5, 8 – 18 and 4 days respectively.

Management strategies:

    i.   Spray application of phosalone 0.07% or endosulfan 0.07 or cypermethrin 0.005%.

3. Spodoptera exigua (Noctuidae: Lepidoptera)

Distribution: Has worldwide distribution which includes Europe, South Africa, America and the oriental region. In India, it is quite widespread, found attacking jute, indigo, Lucerne, lentil, cabbage, maize, cotton and gram.

Nature of damage: The caterpillars, on hatching, gather on the leaf surface, the epidermis of which they eat. At this young stage, they are also in the habit of webbing together either several leaflets or the margin of the same large leaf. At times, these webs give a shabby look to the crop. Within these webs the young larvae live gregariously only for two or three days and thereafter they separate and spread out.

The feeding activity of grown up larva is generally confined to a few morning hours i.e. 9 to 11 a.m. and then again after 4 p.m. or so. They are very voracious and quite large patches of foliages are quickly stripped.

Life history: The adult lays eggs on leaves in clusters of up to 200 eggs each. Each egg is spherical like a poppy-seed in shape and size but with radiating lines. These egg-clusters are often covered with buff-coloured hair which are also present inbetween the eggs. The egg-period ranges between 24 - 36 hours. The adult stage is a typical small noctuid moth with dark –spotted forewings and white hind wings. The colour of the larva is very variable, depending on the crop on which it has been feeding. When full-grown and full fed the caterpillar seeks shelter usually on the soil surface at the base of the plant, under stones or among leaves and such other debris. Also when necessary, a small amount of webbing is produced as a covering and a very rough cocoon is formed with bits of leaf and other material. The larval period is completed in 12 – 14 days. Inside this cocoon the larva pupates and the chrysalis is of the usual noctuid type with a double spine at the tip of the abdomen. The pupal period may be as short as five days and the whole life-cycle can be completed in less than three weeks. But the life-cycle can be very much lengthened, depending upon the environmental temperature and humidity.

Management strategies:

    ii. Collection and destruction of egg masses.

    iii. Spray application of phosalone 0.07% or endosulfan 0.07% or cypermethrin 0.005%.

7.8 PESTS OF PULSES

7.8.1 Gram Pod Borer Helicoverpa Armigera (Hb.) (Noctuidae: Lepidoptera)

Distribution: This polyphagous species is a well known important pest of pulses and cotton in India. It attacks pulse crops such as redgram, Bengal gram, Lab-lab niger, soybean, green gram, black gram and pea. Its other important food plants include safflower, chillies, sorghum, groundnut, tomato and cotton.

Nature of damage: The larvae feed on leaves and bore into pods. As the internal contents of pods are devoured the yield of pulses is considerably reduced.

Life history: The moth has a V-shaped speck on the light brownish fore wings and a dark border on the hind wings. It lays the spherical yellowish eggs singly on tender parts of plants. The full grown larva measures 35 – 45 mm long and is greenish with dark gray lines laterally on the body. While feeding it thrusts its head inside the pod leaving the rest of its body outside. It pupates in an earthen cocoon in the soil. The egg, larval and pupal periods respectively is 2 - 4, 18 - 25 and 6 – 21 days.

Management practices / strategy:

    iv. Setting up of pheromone traps @ 5 traps/ha before the initiation of flowering and collection and destruction of moths caught in the traps.

    v. Spray application of phosalone 0.07 % or endosulfan 0.07% or profenofos 0.05 % or cypermethrin 0.005% or the combination with Bacillus thuringiensis var. kurstaki three times at fortnightly interval commencing from flowering affords protection.

    vi. Dusting of endosulfan 4% or carbaryl 10% dust @ 25 kg/ha once at initiation of flowering controls the pest on Bengal gram.

    vii. Spray application of HaNPV at dusk @ 250 larval equivalent /ha.

7.8.2 Plume moth Exelastis atomosa Walsingham (Pterophoridae: Lepidoptera)

Distribution: It occurs in most of the regions in India where redgram is grown.

Nature of damage: The larva feed on flower buds and seeds of pods by remaining outside and is very characteristic. The pulse crops attacked are red gram and Lab – lab niger.

Life history: The adult is a small greenish-brown plume like moth. These moths lay eggs singly on the tender pods and these eggs hatch into tiny larvae about 1 mm in length within a few days. These young larvae at first scrap their food from the surface of the pod before they cut holes in it. At times they also bore into the unopened flower-buds and feed on the developing anthers. The larvae become full-grown within two to four weeks when they are about 7 mm long. The larva has numerous rosettes of capitate spines and hair all over the body. The pupal period lasts from three days to more than a week, depending on the prevailing temperature.

Management strategies:

i. Spray application of phosalone 0.07% or endosulfan 0.07% or profenofos 0.05%.

7.9 PESTS OF TUBER CROPS

Pests of Potato (Solanum Tuberosum)

7.9.1 Cutworm Agrotis Ipsilon (Hufn.) (Noctuidae: Lepidoptera)

Distribution: The cutworm is an important pest of potato which is widely distributed in India. The larvae are also known to feed on barley, oats, tobacco, pulses, cabbage, beetroot, peas, lady's finger, etc.

Nature of damage: The larvae, which hide during day time in cracks in the soil, become active at dusk, feed on leaves and also cut the tender stems of young and growing plants and thus cause reduction in yield of tubers upto 35 %.

Life history: The adult is a dark brown stout moth with a reddish tinge and has wavy lines and spots on the fore wings. The female lays about 300 eggs in groups of 30 on the under surface of leaves or on moist soil. The larva is dark brown with a red head and when full grown has a greasy appearance which pupates in an earthen cocoon. The egg, larval and pupal periods is respectively 2 – 13, 10 –30 and 10 – 30 days. The total life-cycle lasts from 30 - 68 days depending on weather conditions.

Management strategies:

i. Hand picking and destruction of larvae.
ii. Soil drenching with chlorpyriphos 0.1 emulsion before planting.

7.9.2 Potato Tuber Moth Phthorimaea Operculella Z. (Gelechiidae: Lepidoptera)

Distribution: This is a cosmopolitan pest distributed in different parts of India, which gained entry into India about 70 years ago from Italy. It is a pest of importance on potato both in the field and in storage.

Nature of damage: In storage 30 - 70 % of tubers are damaged by the larvae. In the field the larvae mine into leaves or bore into tender shoots and developing tubers.

Life history: The small dark brown moth lays from 100 - 150 eggs singly on the under surface of leaves or on exposed tubers. Egg period is 3 - 6 days. The yellowish caterpillar with a brown head tunnels into the leaves, stem or tubers and when full grown pupates in a silken cocoon among trash, clods of earth, etc. on the ground or on seams of bags and in crevices in the floor or on walls in storage. Larval period lasts for 5 - 16 days. Hibernation takes place in the larval or pupal stage in some cooler areas of the country. Pupal period is 5 - 7 days.

Management strategies:

- ◉ Earthing up the crop to close the crevices helps in minimizing infestation and in godowns the tubers may be stored in sand.

- ◉ Fumigation of tubers with methyl bromide at 2.5 - 5 kg/1000 cu.m. for 3 hours brings about control of the pest in storage.

- ◉ Spray of phosalone @ 0.07% upon the initial occurrence of the pest in the field. (iv) Release of the parasitoid Copidosoma koehleri @ 1.5 lakhs / ton of stored potato.

8

INSECT PESTS OF HORTICULTURAL CROPS AND ITS MANAGEMENT

8.1 INTRODUCTION

The Horticulture production has become a key driver for economic development in many of the states in the country and it contributes 30.4 per cent to GDP of agriculture. India is globally, second largest producer of fruits and vegetables. Country is the largest producer of mango, banana, coconut, cashew, papaya, pomegranate etc. and also largest producer and exporter of spices. In the foreign trade, export growth of fresh fruits and vegetables in term of value is 14 per cent and of processed fruits and vegetables is 16.27 per cent. Production losses due to pests are around 30 per cent of the total economy of our country. However, study of pest and their management is important in the horticultural crop production.

8.2 PESTS OF VEGETABLES

8.2.1 Pests of Brinjal

I Borers

Shoot and fruit borer : General symptoms of damage are withered terminal shoots, bore holes on shoots plugged with excreta,(Fig. 8) shedding of flower buds, drying of leaves due to boring on petioles by larvae. Larva is pink in colour. Adult is medium sized moth with forewings having black and brown patches and dots. Hind wings are opalescent with black dots(Fig. 7).

Stem borer : Stunted growth, withering and wilting of plants. Bore holes on stem and leaf axils are covered with excreta; Infestation caused by larva. Larva is yellowish or light brown with red head. Moth greyish brown, forewings with transverse lines and white hindwings. **Bud worm** : Larva causes shriveling and shedding of flower buds. It is pale whitish with pink tinge. Adult moth is small with fringed wings

II Leaf feeders

Spotted beetle (or) Hadda beetle : Both grubs and adults feed by scrapping chlorophyll from epidermal layers of leaves which get skeletonized and gradually dry up. Grub is yellowish in colour and stout with spines all over the body. Adult is spherical pale brown and mottled with black spots (6 or 14) on each elytra.

Leaf roller : Leaves are folded from tip to downwards followed by withering and drying up of leaves. Purple brown larva is ornamented with yellow spots and hairs. Adult is with brown forewings and an olive green triangular patch on outer area.

Ash weevils : Adults cause notching of leaf margins. Grubs feed on roots resulting in wilting of plants. Grub is small and apodous. Adults are greenish white with dark lines on elytra or brownish weevil or brown with white spot on elytra or small and light green in color.

III Sap feeders

Leafhopper : Symptom of damage is yellowing of leaves followed by crinkling and downward curling leading to bronzing and hopper burn. Nymph is light green and translucent. Adult is green in colour.

Aphid : Curling and crinkling of leaves, stunted plants with honeydew secretion and sooty mould are the symptoms of damage. Large number of aphids are seen on tender/apical shoots. Nymph is greenish brown or yellow in colour. Adult is yellowish green to dark green in posterior side.

IV Root feeders

Termites: Trinervitermes biformis, Microtermes spp : Nymphs and adults gnaw the roots below the ground level, tunnel upwards through the stems and eat inner tissues. The affected plants wither and dry especially in light soils.

8.2.2 Pests of Tomato

Fruit Borer : Young larva feeds on tender foliage and from fourth instar onwards infests fruits. They make circular holes and thrust only a part of their body inside fruit and eat inner contents.(Fig. 69) Young larva is yellowish white but gradually becomes green. Full-grown larva is apple green in colour with white and dark grey-brown longitudinal lines and sparse short hairs. Adult is light brown and medium sized moth with dull black border.

Stem borer : Stunted growth, withering and wilting of plants, stem and leaf axils covered with excreta covering boreholes are symptoms of infestation caused by the larvae. Larva is yellowish or light brown with red head. Moth is with greyish brown forewings having transverse lines and white hindwings.

II Leaf Feeder

Leaf miner : Leaves are often with serpentine mines followed by drying and dropping of leaves due to infestation. Larva is orange yellowish and apodous. Adult is pale yellow fly.

Tobacco caterpillar : Young larvae scrap the leaves on ventral side. Grown-up caterpillar completely defoliates. Larvae also feed on young fruits. Larva is pale greenish brown with dark markings. Yellow and purplish spots are seen on the submarginal areas. Adult is stout moth with wavy white markings on the brown forewings and white hindwings with a brown patch along its margin.

Green semilooper : Leaves are with holes and skeletonisation and defoliation represent severe damage. Larva is slender, attenuated anteriorly and green in colour with light wavy lines and a broad lateral stripe on either side. Adult is stout moth. Head and thorax are grey in colour. Abdomen is white with basal tufts ferruginous and grey wavy forewings with a slender Y– mark.

Spotted beetles : Both grubs and adults feed by scrapping chlorophyll from epidermal layers of leaves which get skeletonized and gradually dry up. Grub is yellowish in colour and stout with spines all over the body. Adult is spherical pale brown and mottled with black spots (6 or 14) on each elytra.

III Sap Feeders

Green peach aphid : Leaves get curled and crinkled coated with honeydew and sooty mould. Plants remain stunted. Nymphs occur in different 187 colour forms viz.,yellow, green and red. Yellow forms are dominant. Both winged and wingless adults are common.

Fruit sucking moth : Adult sucks the juice by piercing the fruits. Infested fruits will shrink, shrivel, rot and ultimately drop down. Semilooper is with orange blue and yellow spots on its velvety dark speckled body. Stout built moth is with grey and orange coloured wings. Forewing are gray with white patches and a tripod black mark in the center of each(Fig. 70). Hind wings are yellow bearing black patches on the outer margin and curved patch in the middle. The larvae feeds on the leaves of the creeper weed Tinospora vordifolia.

8.2.3 Pests of Bhendi

I . Borers

Stem weevil : The grub causes gall like swellings on the stem near the base. Grub is white and apodous. Adults feed on leaves, buds and tender terminal shoots. Grub is creamy yellow and apodous. Adult is dark greyish brown with pale cross bands on elytra.

Shoot weevil : Grubs bore into stem and petioles causing gall like swellings. Adults feed on leaves, buds and tender terminal shoots. Grub is creamy yellow and apodous. Adult is dark greyish brown with pale cross bands on elytra.

Shoot and fruit borer: Symptom of attack is withering and drying of tender shoots in the early stage. Larva bores into flowers and flower buds causing withering and dropping of the same. Fruits with bore holes are seen often and sometimes deformed. Larva is stout , spindle shaped, dark 188 brown in colour and with white patches on the dorsum. Earias vitella : small buff colored. Forewings with a wedge shaped green patch in the middle. E. insulana: entire forewing is green.

Stem fly : The maggot bores into tender shoots and petiole of leaves resulting in drying of leaves and seedlings. Maggot is yellow in colour. Adult is a small black fly.

Fruit borer : Young larva feeds on tender foliage and from fourth instar onwards attack fruits. They bore circular holes and thrust only a part of their body inside fruit and eat inner contents. Freshly hatched larva is yellowish white but gradually become green. Full-grown larva is apple green in colour with white and dark grey-brown longitudinal lines and sparse short hairs. Adult is light brown and medium sized moth with dull black border.

III Leaf

Feeders Leaf roller : Young larvae feed on the epidermis, roll the leaves, feed within and eat away the rolled portions. Larva is bright green with dark head and prothoracic shield. Moth is with yellowish fore and hindwings with brown lines and distinct markings.

Semilooper : The caterpillar completely feeds on the leaves (defoliation). Anomis flava: Larva is green in colour with 5 white longitudinal lines. Adult is brown and medium sized moth. Acontia (=Xanthodes) graellsii: Larva is green in colour with horse-shoe-shaped black markings on each segment. Moth is yellowish with black markings 189 all over the wings. Tarache nitidula: Larva is green in colour and resembles bird"s droppings. Adult is white with grey markings.

Tobacco caterpillar: Young larvae scrap the leaves on ventral side. Grown-up caterpillar completely defoliates. Larvae also feed on young fruits. Larva is pale greenish brown with dark markings. Yellow and purplish spots are seen on the submarginal areas. Adult is stout moth with wavy white markings on the brown forewings and white hindwings are having a brown patch along its margin.

8.2.4 Pests of Cucurbits

Fruit flies : The maggots feed on the pulp of the fruits and the symptoms of damage include oozing of resinous fluid from fruits, distorted and malformed fruits (Fig. 71) premature dropping of fruits and unit for consumption. Maggot is white and apodous. Adult is with hyaline wings or brownish body with brown oval spot on either side of 3rd tergite.

Snake gourd stem weevil : Grub bores into the stem/petiole and causes withering of leaves. Adult is small black weevil and feeds on leaves.

Stem gall fly : Maggot bores inside the distal shoot and induces galls. Adult is slender and dark brown mosquito like fly.

Stem borer : Larva bores into the stem of snake-gourd and produces gall. Adult is dark brown moth with transparent wings.

Leaf miner : Leaves are often with serpentine mines followed by drying and dropping of leaves due to infestation. Larva is orange yellowish and apodous. Adult is pale yellow fly.

Snake gourd semilooper : Larva cuts the edges of leaf lamina, folds it over the leaf and feeds from within the leaf roll. Larva is whitish green and the body is with black warts, off-white longitudinal stripes and a hump on its anal segment. Stout dark brown adult has shiny brown forewings.

Pumpkin caterpillar : The caterpillars lacerate and feed on chlorophyll of foliage; later fold and web the leaves together and feed within. They may also damage ovaries of flowers and boring into young developing fruits. Larva is elongate, bright green with two narrow longitudinal stripes dorsally. Adults are medium sized, wings are white and transparent with broad brown margin. Female has tuft of orange hairs at the anal end.

Pumpkin beetle: Grubs feed on the roots, stem and fruits that spread over the soil. Adults feed on leaf and flower. Grub is creamy yellow. Adult is grey black with black or blue colour with glistening yellow red border.

8.2.5 Pests of Crucifers

I Leaf Feeders

Diamond back moth : Young caterpillars cause small yellow mines followed by scrapping of epidermal leaf tissues producing typical whitish patches. Full-grown larvae bite holes in the leaves. Larva is pale yellowish green in color, pointed at both ends with fine erect black hairs scattered over the body. Adult is small, green brown with pale whitish narrow wings. At rest a dorsal median patch of 3 diamonds shaped 191 yellowish white spots are clearly visible by joining both forewings. Hindwings have a fringe of long fine hairs(Fig. 72).

Cabbage borer : Larvae web the leaves and bore into the stem, stalk or leaf veins. Larva is pale whitish brown with 4-5 purplish brown longitudinal lines. Adult is pale greyish brown with 4-5 purplish brown longitudinal lines. Adult is pale greyish brown moth with forewings having grey wavy lines. Hindwings are pale dusty.

Leaf webber : Young larvae feed gregariously on leaves, later web together the leaves and feed. Larva is with red head, brown longitudinal stripes and rows of tubercles with short hairs on its pale violaccous body. Adult is small with brown forewings having distinct wavy spots. Hindwings are semi-hyaline.

Cabbage semilooper : Damaged leaves are with holes initially and the severe damage is represented by skeletonization. Larva is green color with light wavy lines and broad lateral stripes on either side. Adult is stout moth. Head and thorax are grey in colour and the abdomen is white with basal tufts. Head and thorax are grey in colour and the abdomen is white with basal tufts. Grey wavy forewings are with a slender „y" mark. Cabbage butterflies : The caterpillar feeds on leafy vegetation irregularly (defoliation). Sometimes bores into the heads of cabbage. Larva is velvety bluish green in colour with yellow dorsal and lateral stripes are covered with black hair. Adult is with snow white forewings and black apical spots; hind wings are pure white. 192

Tobacco caterpillar : The caterpillar damages leaves and heads of Cabbage, Cauliflower, Radish and . . Larva is pale greenish brown with dark markings. Yellow and purplish spots are seen on the submarginal areas. Adult is stout moth with wavy white markings on the brown forewings and white hindwings with a brown patch along its margin.

Mustard sawfly : Caterpillar like grubs nibble the tender margins of tender leaves and later bite holes on the leaves. Adult is with dark head and thorax, orange coloured abdomen and smoky wings with black veins. Female has a strong saw-like ovipositor.

II Sap Feeders

Thrips : Nymphs and adults suck the sap from leaves. Nymph is pale yellow and the adult has fringed wings.

Mustard aphid : Nymphs and adults suck the sap from the under surface of the leaves. Nymph is light yellowish green and adult is darker than nymph.

Cabbage aphid : Nymphs and adults cause crinkling and cupping of distorted primordia. White cast skins are present at the base of the plant. Adult is yellowish green with wavy white filament over the body.

Painted bug : Nymphs and adults desap the leaves, shoots and pods. Adults are small black bugs with red and yellow lines.

8.2.6 Pests of Moringa

Moringa

Bud worm : Larvae bore into flower buds and causes shedding. Larva is dirty brown with mid-dorsal stripe and black head with prothoracic 193 shield. Adult is small with dark brown forewings and white hindwings with a brown border.

Bud midge : Feeds on internal content of flower bud and causes shedding. Adult is small brownish fly.

Leaf caterpillar : Larva remains in a silken web in the undersurface of leaf and feeds on the leaflets reducing them into papery leaf. Larva is with brown head and without prothoracic shield. Adult is bigger than bud worm.

Moringa hairy caterpillar : Larvae are seen in groups in tree trunks and feed gregariously, scrap the bark and gnaw the foliage resulting in defoliation of tree. Larva is brown and hairy. Adult is large sized, uniformly light yellowish brown in colour with faint lines on wings.

Black hairy caterpillar : Caterpillars feed on leaf lamina initially by scrapping epidermal layers and later by cutting the blades.

Pod fly / fruit fly : Severe infestation results in drying of fruits from tip. Gummy exudate oozes from infested fruits. Adult is small yellowish fly with red eyes.

Bark caterpillar : Zigzag galleries and silken webbed masses comprising of chewed material and excreta of larvae are seen. Larva is stout and dirty brown. Adult is pale brown; forewings with brown spots and streaks.

Stemborer : Grub causes zigzag burrows beneath the bark, which results in death of the branch, or stem. Adult feeds on bark of the young petiole 194 and twigs. Grub is stout and yellowish. Adult is large sized beetle with yellowish brown elytra.

8.2.7 Pests of Potato

I Leaf Feeders

Common cutworm : Young larvae feed on leaves and the grown up larvae cut the stem at collar region. Larva is black colored with brown head. Adult forewing is grey with spot like markings. Hind wing is dull white.

Black cutworm : Damage as in common cutworm. Larva is black with pale mid dorsal stripes. Adult forewing is pale brown with dark purplish brown and hind wing is with brown tinge.

Spotted beetle : Both grubs and adults feed by scrapping chlorophyll from epidermal layers of leaves which get skeletonized and gradually dry up. Grub is yellowish in color and stout with spines all over the body. Adult is spherical pale brown and mottled with black spots (6 or 14) on each elytron.

Bihar hairy caterpillar : Young larvae feed gregariously and skeletonize the leaves. Later instars defoliate completely. Larva is stout with seven orange transverse lines with tuft of yellow hairs, which are dark at both ends. Adult is crimson colored, body with black dots and black antenna. Wings are pinkish with black spots. 195

II Borers

Shoot and fruit borer : General symptoms of damage are withered terminal shoots, bore holes on fruits plugged with excreta, shedding of flower buds, drying of leaves due to boring on petioles by larvae. Larvae are pink in color. Adult is medium sized moth with forewing having black and brown patches and dots. Hind wings are opalescent with black dots.

Potato tuber moth : It is a pest of field and storage. Larva tunnels into foliage, stem and tubers, which leads to loss of leaf tissue, death of growing points and weakening or breaking of stems.(Fig. 74) In tubers, irregularly shaped galleries are formed near tuber eyes. Larva is white to yellow or greenish turns red at pupation. Moth is small with silvery body. Forewing is grey-brown with minute dark spots and has a narrow fringe of hairs. Hindwings are dirty white (Fig. 73).

Root grubs : Grubs feed on roots and tubers. Adult feed on foliage during night. Damage is more during autumn. Grub is „C" shaped with orange head. Adult is brown beetle with pale prothorax.

8.2.8 Pests of Sweet Potato

I Borers Sweet potato weevil : Grubs bore into stem and feed on soft tissues. Grubs and adults bore into tubers both in field and in godowns. Occasionally adults feed on stem and leaves as well. Grub is fattish, apodous and pale yellowish white in colour. Adult is ant like, slender bodied having elongated snout, bluish brown head with non-geniculate antennae, bright red thorax, brownish legs and red abdomen. 196

Tuber borer : Caterpillars bore inside the tubers and feed the starchy material. The adults are grayish brown; forewings are mottled with fine specks and grayish lines and black spots.

Stem or vine borer : Caterpillar bores into vines (stem) often killing the branch. Larva is stout and whitish in colour. The moth is yellow with dark wavy lines.

II Leaf Feeders

Leaf roller : Tiny larvae scrape the tender surface tissue of leaves and feed in beneath the thin webbings. Larva folds single leaf longitudinally and feed on green tissues

Tortoise beetles : Grubs and adults bite holes on leaves. Grub is flattened yellowish green with spiny processes covering the body. It has a raised anal portion with which it covers its back with excreta and carries the skin on its back. Adult beetle is medium sized with colour variation according to species. A. miliaris: Broad oval shaped, brownish red in colour with black dots.

C. circumdata: Beetle with green crescent –like mark in the middle

C. bipunctata: Small metallic green with six black spots on elytra.

III Root Feeder

White grub : Grubs feed on roots and tubers and adults feed on leaves. Adult is chestnut colored beetle with glistening pubescence.

8.2.9 Pests of Tapioca

Cassava scale : Nymphs and adult desap the plant and cause stunting and death. White elongate scales are present on stem.

Whitefly : Nymphs and adults cause chlorotic spots by sucking cell sap from leaves and then yellowing and drying of leaves. Nymph is greenish and oval in outline. Adult is with yellow body covered with white waxy bloom.

Thrips : Nymphs and adults cause silvery patches on leaves. Nymph is reddish in colour. Adult is dark brown or black.

8.2.10 Pests of Chillies Chillies

Stem borer : Stunted growth, withering and wilting of plants, stem and leaf axils covered with excreta covering bore holes are the infestation caused by the larvae. Larva is yellowish or light brown with red head. Moth is with greyish brown forewings having transverse lines and white hindwings.

Chilli thrips : Leaves become crinkled, curled upward and shed. Buds become brittle and drop down. Plants get stunted and bronzed. Nymphs and adults are tiny slender, fragile and yellowish straw in colour.

Green peach aphid : Leaves get curled and crinkled coated with honeydew and sooty mould. Plants remain stunted. Adult is mostly yellow in colour.

Tobacco caterpillar : Young larvae scrap the leaves on ventral side. Grown-up caterpillar completely defoliates. Larvae also feed on young fruits. Larva is pale greenish brown with dark markings. Yellow and purplish spots are seen on the submarginal areas. Adult is stout moth with wavy white markings on the brown forewings and white hindwings are having a brown patch along its margin.

Cut worm : The greasy cut worms come out during night and curt the seedlings at ground level and eat tender leaves.

Fruit borer : Young larvae feed on tender foliage and from fourth instar onwards attacks fruits. They bore circular holes and thrust only a part of their body inside fruit and eat inner contents. Freshly hatched larva is yellowish white but gradually become green. Full-grown larva is apple green in colour with white and dark grey-brown longitudinal lines and sparse short hairs. Adult is light brown and medium sized moth with dull black border.

Muranai mite : Sudden curling and crinkling of leaves followed by blister patches are initial symptoms. Plants are severely attacked, stop growing and die. Adult is tiny, oval, glossy or whitish mites.

Eriophyid mite : These mites infest tender shoot cause rusting, leaf size reduction and shedding of flowers.

8.2.11 Pests of Cardamom

I Borer

Shoot, panicle and capsule borer : Pseudostem with bore holes plugged with excreta, dead heart, panicles and spikes dry up above the point of infestation and empty capsules are the damages caused by the caterpillar. Larva is pale greenish with pinkish tinge and fine hairs with dark head. Adult is medium sized moth, pale yellowish with small black spots on the wings.

Rhizome weevil : The grubs break the tillers at the base, which results in rotting, falling down and drying of clumps. Grub is glossy white with light brown head. Adult are brown weevil with 3 lies on the pronotum and 3 black spots on each elytron.

8.2.12 Pests of Pepper

I Borers

Pollu beetle : Affected berries are with exit holes, dry up later, turn dark and hollow and crumble when pressed. Irregular feeding holes are seen on leaves. Young grub is with transparent body and grown up is yellow or brownish. Adult is oblong beetle with broad body and shiny black elytra and enlarged hind femur.

Top shoot borer : The infestation results in drying of terminal shoots. Larva is greyish green and 12 –14mm long. Adult is a tiny moth. Forewing is black and distal half is red. Hindwing is greyish.

8.2.13 Pests of Betelvine

Aphid : Nymphs and adults desap the tender shoots and leaves and result in crinkling and curling of leaves and drying. 200

Whitefly : Nymphs and adults cause chlorotic spots by sucking cell sap from leaves resulting yellowing and drying of leaves. Nymph is greenish and oval in shape. Adult is with yellow body covered with white waxy bloom.

Scale : Infests the leaves, petiole and main vines. Infested leaves lose their colour, exhibit warty appearance., crinkle and dry up ulimately. Vines present a sickly appearance and wilt in due course. Adult is dark brown or light brown scale.

8.2.14 Pests of Mango

I Pests of Inflorescence/Fruit

Mango hoppers : Nymphs and adults cause withering and shedding of flower buds and flowers. Presence of small drops of honeydew on lower leaves followed by development sooty mould. Clicking sound due to movement of jassids amidst leaves is a common phenomenon.

I. niveosparsus - Three spots on scutellum and white band across the wing.

I. clypealis - Two spots on scutellum and dark spots on the vertex

A. atkinsoni - Two spots on scutellum.

Aphid : The infestation results in drying of inflorescence and tender shoots and appearance of sooty mould. Aphids are brown coloured.

Flower webber : Larvae web the inflorescence and tunnel the stalks. Larva is greenish yellow light brown head and prothoracic shield. Adult female moth is with grey wings and male is with purplish pink wings. 201

Gall midges : Procystiphora mangiferae - Causes malformation of flowers and droppings of flower. Maggot and adult are orange coloured. Dasineura amaramanjarae - Causes damage to flower buds and dropping of bud. Erosomyia mangiferae - Results in stunting and malformation of inflorescence. Maggot is yellowish.

Fruit fly : Semi-ripe fruits are with decayed spots and droppings of fruits. Maggot is yellowish. Adult fly is light brown with transparent wings.

Nut weevil : The infestation results in dropping of fruits at marble stage and tunnelled cotyledons. Ovipositional injuries and eggs are seen on marble sized fruits. Grub is fleshy, yellowish and apodous. Adult is brownish with short snout and papillate scales (Fig. 75).

II Leaf Feeders Shoot Webber :

Larvae cause webbing of terminal leaves and defoliation. Larva is pale green with brown head and prothoracic shield. Adult is brownish moth with wavy lines on forewings.

Castor slug : Larva irregularly feeds on the leaves and causes defoliation. Larva is slug like, ventrally flat, greenish body with white lines and four rows of spiny scoli tipped red or black. Adult is green moth with a brown band at the base of forewings. 202

III Sucking Pests

Whitefly : Nymphs and adults cause yellowing of leaves in patches and the presence of white flies on the ventral side of leaves. Nymphs are greyish white, found in-groups. Adult is dull white in colour.

Scale insect : Nymphs and adults cause yellowing of leaves. They are white elongate hard scale.

Mealy bug : Severe infestation results in drying of leaves and inflorescence. Nymphs and adults are pinkish and undergo diapause in soil during winter.

Eriophyid mite : This worm like mites are found in growing tips, sucking the sap and injecting toxic substances, kill the buds and cause resetting of shoot.

IV Borers

Stem borer : The grub causes drying of terminal shoots in early stage of attack. Wilting of whole tree damage occurs at the main stem. Grub is linear, fleshy and apodous. Adult is greyish beetle with two pink dots and lateral spine on the thorax.

Shoot borer : Larvae bore through the downwards from the growing tip to a depth of 5 or 6 inches. Whole seedling remains stunted with individual twigs showing a peculiar terminal bunchy appearance. Larva is dark pink with conspicuous dark brown prothoracic shield. Adult is greyish moth with dark grey wings having wavy designs.

8.2.15 Pests Of Citrus

I Internal Feeders

Orange borer : The grubs cause drying of terminal shoots in the early 203 stages, followed by wilting of thicker branches and main stem. Grub is creamy white with flat head. Adult is dull metallic green to dark violet or shiny blue beetle with yellow band across the middle of the elytra.

Citrus leaf miner : The infestation by the larva results in leaves with serpentine mines and distortion of the leaf lamina. Larva is minute, reddish or yellowish and apodous. Adult is minute moth with a black spot at the tip of the forewing (Fig. 76).

II Leaf Feeder

Citrus butterfly : The larva causes defoliation of tender leaves. Larva in its early stage resembles bird dropping. Grown up larva is cylindrical, stout and green with brown lateral oblique bands. Adult is dark brown swallow tail butterfly with numerous yellow markings.

III Sap Feeders

Fruit sucking moths : Adult moths pierce the fruit and suck the juice resulting in rotting at the feeding site and dropping fruit. Larva is semilooper with orange blue and yellow spots on its velvety dark speckled body, which feeds on the weed host. Adult is stout-built moth with grey and orange coloured wings. There are 3 black spots on the forewings.

Otheris fullonica - Presence of tripod black marks in the forewing and curved marking in the hindwing. O.ancilla - Presence of white band in the middle forewing. 204

Psyllid : Nymphs and adults infest terminal tender twigs and desap causing curling and drying of twigs. Transmit citrus greening virus. Nymph s are orange in color; adults are brownish males are shorter than female. Wings are memberaneous and semitransparent, wings extend beyond the body.

Citrus whitefly : Nymphs and adults suck sap from leaves causing curling over and fall off. Nymph is pale yellow with purple eyes

Black fly : The symptoms of damage are yellowing of leaves in the early stage of attack followed by honeydew deposition on the lower leaves and sooty mould of development. Severe infestation leads to defoliation. Nymph is shiny black scale like and spiny with white markings at the edges.

Scale insects : Suck sap from branches, inject toxic substances. Females are light grey in color; males are smaller than females.

Rust mite : Feeding by adults and nymphs causes silvery, scaly of rusty to black discolouration on the fruits. The affected fruits are smaller and the rind of injured fruits is thicker.

8.2.16 Pests of Sapota

Chickoo moth or leaf webber : Leaves are webbed together in a bunch and the chlorophyll scrapped by the larva. Cluster of dried leaves is hanging from the webbed shoots (Fig. 77). Flower buds and tender fruits are bored, become withered and shed. Larva is pinkish in colour with 205 three dorso-lateral brown stripes on each side. Adult moth is greyish with hairy brown forewings or black spots and semi hyaline hindwings.

Budworm : Floral buds and flowers are webbed together and shed. Larva is small, slender, pinkish brown in colour with black head and yellowish brown prothoracic shield. Adult is grey coloured moth with black patch on wings.

Fruit fly : Semi–ripe fruit show decayed spots and fruits drop later. Maggot is yellowish. Adult fly is light brown with transparent wings.

Hairy caterpillar : Larva feeds on leaves irregularly and causes defoliation. Larva is greyish brown, stout and hairy. Adult is stout greyish brown moth. Male is with pectinate antenna and chocolate brown patch in the middle of forewings. Female is bigger in size than male and has wavy transverse bands on wings.

8.2.17 Pests Of Guava

Fruit borer : Infected fruits are with boreholes plugged with anal segment of the larva. Severe infestation results in fruit rotting and dropping. Larva is dirty dark brown, short and stout built covered with short hairs. Adult is bluish brown butterfly. Female is with „V" shaped patch on forewing. Fruits are with boreholes. Adult is metallic red coloured butterfly.

Tea mosquito bug : Corky scab formation on fruits is the symptom of damage. The infestations caused by the nymphs and adults caused by the nymphs and adults include inflorescence blight, terminal drying of young shoots and water soaked lesions followed by brownish spots at the 206 feeding sites (Fig. 78). Nymphs and adults are reddish brown, elongate bugs with black head, red thorax and black and white abdomen.

8.2.18 Pests Of Pomegranate

Fruit borer : Infested fruits are with bore holes plugged with anal segment of the larva. Severe infestation results in fruit rotting and dropping. Larva is dirty dark brown, short and stout built covered with short hairs. Adult is bluish brown butterfly. Female is with „V" shaped patch on forewing.

Fruit fly : Rotting of fruit is the symptom of infestation. The maggots feed on the pulp of the fruits and the symptoms of damage include of brown resinuous fluid from fruits, distorted and malformed fruits premature dropping of fruits and unfit for consumption. Maggot is white and apodous. Adult is with hyaline wings or brownish with pale yellow band on 3rd tergite.

Shoot and fruit borer : Larvae make holes on fruits. Larva is pale greenish with pinkish tinge and fine hairs with dark head and prothoracic shield. Adult is medium sized and pale yellowish moth with small black spots on the wings.

Mealy bug : Cluster of white mealy bugs on the lower-side of the older plants cause yellowing and drying of leaves. Adults are small, oval, soft bodied and covered with white mealy wax. 207

8.2.19 Pests of Banana

I Borers

Rhizome weevil : The grub causes death of unopened pipe and withering of outer leaves. Grubs bore into the rhizome and cause death of the plants (Fig. 79). Grub is apodous and yellowish white with red head. Adult is dark coloured weevil.

Pseudostem borer : The grub makes bore holes and tunnels in the pseudostem and causes wilting of the plant. Grub is apodous and creamy white with dark head. Adult is robust reddish brown and black weevil.

II Sap Feeders

Banana aphid : Nymphs and adults are vectors of bunchy top disease. They are seen in colonies on leaf axils and pseudostem. Nymphs and adults are dark

Banana aphid : Nymphs and adults are vectors of bunchy top disease. They are seen in colonies on leaf axils and pseudostem. Nymphs and adults are dark in colour. Winged adults are with black veined wings.

Tingid : The infested leaves are with greyish yellow spots and stunted growth. Presence of white, transparent adults is dull coloured nymphs on the lower surface of leaves. Nymphs and adult are dull coloured bugs with transparent shiny lace-like reticulate wings.

Thrips : Leaf thrips - Cause yellowing of leaves . Adults are with fringed wings. Fruit rust thrips - Cause leaf yellowing and rusty growth over fruit. Adult is yellowish white with shaded wings. Flower thrips - Cause corky scab on fruits and flowers. 208

8.2.20 Pests Of Cashew

I Borers

Cashew tree borer : The grub by internal tunnelling causes wilting of branches and then the tree as a whole (Fig. 80). It also infests trunk and root. Grub is elongated, creamy white brown head. Adult is reddish brown longicorn beetle.

Bark feeder : Zig-zag galleries and silk webbed masses comprising of chewed material and excreta of larvae are seen. Larva is stout and dirty brown. Adult is pale brown with forewings having brown spots and streaks.

Apple borer : Presence of bore holes on the tender cashew (or) apple. Larva is dark pink in colour. Adult is medium sized moth with dark forewings and pale hindwings.

II Inflorescence Feeders

Shoot and blossom borer : The larva causes webbing of tender leaves and inflorescence. Larva is reddish brown with yellow and pink lines. Adult male is dark fuscous. Female is pale and olive green.

Shoot tip and inflorescence caterpillar : The infestation results in webbing of terminal leaves and inflorescence and boring of shoot tip. Larva is yellowish brown. Adult is a dark and tiny moth. 209

Tea mosquito bug : The infestations by the nymphs and adults include inflorescence blight, terminal drying of young shoots and water soaked lesions followed by brownish spots at the feeding sites.

III Leaf Feeders

Leaf miner : Mining of tender leaves in whitish blotches is the symptom of damage. Larva is reddish brown and minute. Adult is silvery grey moth with fringes of hairs on the wing margins.

Wild silk moth : The larvae feed on leaves, which results in complete defoliation. The infestation is indicated by the presence of golden coloured pupae on the trunk. Larva is stout, dark brown with prominent warts all over the body. Adult is pale yellowish or reddish brown moth with three clear moths on forewings.

Hairy caterpillar : Larva feeds on leaves irregularly and causes defoliation. Larva is greyish brown, stout and hairy. Adult is stout greyish brown moth.

Leaf miner : Mining of tender leaves in whitish blotches is the symptom of damage. Larva is reddish brown and minute. Adult is silvery grey moth with fringes of hairs on the wing margins.

Wild silk moth : The larvae feed on leaves, which results in complete defoliation. The infestation is indicated by the presence of golden coloured pupae on the trunk. Larva is stout, dark brown with prominent warts all over the body. Adult is pale yellowish or reddish brown moth with three clear moths on forewings. 210

Hairy caterpillar : Larva feeds on leaves irregularly and causes defoliation. Larva is greyish brown, stout and hairy. Adult is stout greyish brown moth. Male is with pectinate antenna and chocolate brown patch in the middle of forewings. Female is bigger in size than male and has wavy transverse bands on wings.

Leaf twisting weevil : The grub rolls leaf terminal, results in drying. Grub is yellowish and apodous. Adult is reddish brown weevil.

Looper : The larva damages the leaf margins. It is a green looper. Adult is green with grey brown markings.

IV Sap Feeders

Aphids : The infestation results in drying of inflorescence and tender shoots and appearance of sooty mould. Aphids are brown coloured.

Red banded thrips : The damage results in crinkling, discolouration and leaf drop. Nymph is greenish yellow with red cross band across first two and last abdominal segments.

Thrips : This species causes silvery white patches on leaves with excreta. Yellowing and withering are due to severe infestation. Nymph is reddish in colour. Adult female is dark brown with yellow legs and antennae. Male is with yellow abdomen. 211

8.2.21 Pests of Grapevine

Stem girdler : The grubs and adults cause wilting of branches and then the entire vine. Adult is medium sized and grey coloured with a white spot in the centre of each elytron.

Chafer beetle : The adults cause complete defoliation of the leaves. Adults are brown coloured beetles.

Ground beetle: The adults cause defoliation of the leaves. Adults are brown coloured beetles.

Flea beetle : The adults bite small holes on tender leaves and the root is damaged y the grubs (Fig. 81). Adult is reddish brown, shiny beetle with six spots on elytra.

Leaf roller : The larva causes rolling of leaves. Larva is pale green with short hairs. Adult is brownish moth with wavy line.

Thrips : Leaf thrips - Cause yellowing of leaves. Adults are with fringed wings. Fruit rust thrips - Cause leaf yellowing and rusty growth over fruit. Adult is yellowish white. Flower thrips - Cause corky scab on fruits and flowers.

Blackfly : Yellowing of leaves is the symptom of damage caused by the nymphs and adults. Nymph is oval in shape, scale like fringes. Adult is minute, delicate insect.

Mealy bugs : Nymphs and adults cause crinkling and yellowing of leaves and rotting of berries. 212

Berry plume moth : Larvae cause feeding injury on berries. Larva is small, pale green or pink with median red line. Adult is a small moth.

Castor semilooper : Adult causes fruit rotting and dropping. Larva in varying shades of colour. Head is black with black and red spot on the 3rd abdominal segment and red tubercles on the anal region. Adult is pale reddish brown with black hind wing with a median white and 3 large white spots on the outer margin.

8.2.22 Pests of Coconut

Rhinoceros beetle : Damage is caused by adult beetles which burrow the leaf sheaths near the crown and cut across the leaf in the folded condition. The damaged leaves show characteristic clippings or holes in the leaflets. The infestation will result in stunting of trees and death of growing point. Adult beetle is stout, black, about 5 cm long and has a long horn projecting dorsally from the head in male, a short horn in female. The grubs feed on decaying vegetable matter and in manure pits at a depth of 5-30 cm.(Fig. 82) damaged leaves show characteristic clippings or holes in the leaflets. The infestation will result in stunting of trees and death of growing point. Adult beetle is stout, black, about 5 cm long and has a long horn projecting dorsally from the head in male, a short horn in female. The grubs feed on decaying vegetable matter and in manure pits at a depth of 5-30 cm.(Fig. 82)

Red palm weevil : A few small holes with protruding chewed fibrous material and oozing out of a brown liquid from such holes indicate early 213 infestation. In advanced stage of attack the central shoot shows sign of wilting and on large mass of grubs, pupae and adults are seen inside trunk. The reddish brown weevil has six dark spots on thorax and in the male a conspicuous long snout has a tuft of hairs.

Black headed caterpillar : The larvae live on the undersurface of leaflets within galleries of silk and frass material and feed by scrapping the green matter. The caterpillar is greenish brown with dark brown head and prothorax, and reddish mesothorax.

White grub : The grubs feed on roots and cause stunting and delayed flowering. Adult beetles emerge after monsoon showers.

Termite : Termites damage coconut seedlings.

Scale insect : The undersurface of leaflets is infested by scale insects in large numbers causing yellowing in patches.

Lace wing bug : The nymphs and adults of the lacewing bug feed by sucking the sap from the undersurface of leaflets causing white spots on the upper surface.

Perianth mite : The mite infest and develop on the meristematic tissues under the perianth. Initial symptoms exhibit as triangular pale white or yellow patches close to the perianth. Continuous feeding results in necrosis of tissues leading to formation of brown color patches, longitudinal fissures and splits on the outer surface of the husk; oozing of brown gummy exudation; reduced nut size and copra content and malformation of nuts. The mite is vermiform, elongate body with 2 pairs 214 of legs in the anterior and of the body ; head with piercing and sucking mouth parts.

8.2.23 Pests of Coffee

I Borers

White borer : Presence of ridges on the stem, yellowing of leaves, wilting of branches and occasional drying of plants are the symptoms caused by the grub. Grub is white or yellowish, anterior and broader and tapering towards tail end. Adult is black, elongate beetle with grey pubescence on the head, thorax and elytra and characteristics white markings on the elytra.(Fig. 83)

Red borer : The larvae cause wilting of branches or plant. Boreholes often are plugged with excreta at the base of the plant. Larva is orange red and smooth, Adult is with dirty white bands and black or steel blue spots on the wings. Shot hole borer : Adult and grub make small holes on the under surface of young succulent branches between nodes which result in withered and dead branches with shot holes. Grub is milky –white and apodous. Adult is reddish brown to dark brown beetle with a short cylindrical body.

Coffee bean beetle : Infested berries are with small holes, black in colour and shrunken. Grub is milky-white and apodous. Adult is pale grey, elongate oval and slightly flattened tapering anteriorly. Entire body is clothed with hairs. 215

Berry borer : Infestation by the grubs and adults results in dropping of tender berries. There are many small round holes in the nodal region of developed berry. Damage is often caused to endosperm by making small galleries near the main tunnel. Female adults tunnel into berries. Grub feeds on beans. Grub is white in colour. Adult is black beetle and the males are wingless.

II Leaf Feeder

Leaf miner : The maggots often mine the leaves. Maggot is small and apodous. Adult is very small and **brown coloured fly.**

8.2.24 Pests of Tea

I Leaf Feeders

Looper : Larva causes defoliation of leaves. Larva is grey or dark green in colour. Adult is straw coloured moth. Wings are grey with light brown markings and wavy lines.

Bunch caterpillars : The larvae cause defoliation of leaves. Larva is smooth and hairless and grey in colour with brown patches. Adult is golden brown moth.

Lobster caterpillar : Defoliation of leaves is the symptom of damage by the larva. Larva is brown with white band and elongated legs.

White grub : Grubs feed on roots and rootlets resulting in drying of young plants. Adults are leaf feeders. Grub is fleshy and „C" shaped. Adult is a brown coloured beetle. 216

Tea tortrix : Caterpillar makes leaf nest by webbing the leaves. Adult is greenish with black prothorax or brown coloured bell shaped moth. Male is smaller than female. Larva is greenish with black prothorax.

Tea leaf roller : Second instar larva mines the tender and reaches leaf margin. Fourth instar larva rolls the leaves from the tip downwards. Larva is yellowish. Adult is microlepidoptera. Antenna is longer than the body with golden iridescent patches in forewing and abdomen.

Flush worm : Larvae web the tender leaves enclosing the bud; feed on upper epidermis of leaves and apical portion of the bud. Larva is brown colored and 1 cm long; adult is less than 1 cm in size blackish brown in color.

Nettle grubs : The caterpillars are the nuisance to the workers because of their stinging hairs besides scrapping the leaves.

Faggot worm : Larva defoliates order leaves and also feeds on bark. Adult male is reddish brown and winged. The female is wingless and grub like

II Sap Feeders

Red spider mite : Feeding by nymphs and adults causes the leaves to become bronzed dried and crumpled. Nymph and adult are brick red in colour and rounded.

Scarlet mite : The infestation results in brownish leaves. Large number of miles are seen near the petiole and along in the midrib. Nymph and adult are orange and flattened ovate mite. 217

Purple mite : This species causes brown or coppery brown or smoky discoloration of leaves. Adult mite is dark purple to pink in colour with characteristic white sides running along the back.

Pink mite or orange mite : Continuous desaping causes the leaves turn pale and curl upward. Under severe infestation, leaves become leathery and brown. Damages are often to top 10-15 cm tender leaves. Assam type of tea is susceptible. Nymph and adult are microscopic orange coloured mite and its body is carrot shaped with two pairs of legs.

Yellow mite : The damage is restricted to top two to three leaves and the bud. Leaves become rough and brittle. Corky line or patches appear on the lamina Internode gets shortened, stunted and deformed. Mites are pale yellow in colour. Male is shorter than female with tapering abdomen and a sucker. Fourth pair of legs is provided with a curved tooth and a pair of whips. They carry female on their back. Female is bigger than male with two pairs of whip.

Thrips : Opened leaves show a parallel brown streaks on either side of midrib. Leaf surface becomes uneven. Nymph is creamy white and adult is with fringed wings and brown abdomen.

Tea mosquito bug : Brownish patches are seen in the tender shoots, buds and stem. Curling of leaves and drying of shoots are caused due to severe attack. Adult is black and red elongated insect with long legs and a dorsal process on the scutellum (Fig. 84). 218

Scale : There are many hemispherical brown scales seen along the midrib and tender stem followed by sooty mould on lower leaves. Vegetatively propagated clones are susceptible. Nymph is white, adult is winged and female is sedentary.

8.4 PEST MANAGEMENT

8.4.1 Pest Management in Brinjal

Shoot and fruit borer

- Collection and destruction of infested plant parts like shoots, buds and fruits.
- Avoid ratooning to minimize shoot and fruit borer infestation. 222
- Spray anyone of the following twice at 30 days after planting at fortnightly interval.
- Quinalphos 25 EC 2 ml/lit + NO 2 ml/lit + Teepol 1 ml/lit.
- Neem Seed Kernel Extract (NSKE) 5% (50 g/lit).
- Avoid synthetic pyrethroids.
- Growing resistant varieties like Pusa Purple cluster, Arka Kusmak, Doli 5 etc.,

Ash weevil

Apply carbofuran 3G @ 15 kg/ha, 15 days after planting.

Aphid

- Release the first instar grubs of Chrysoperla carnea @ 10,000/ha.
- Spray methyl demeton 25 EC or dimethoate 30 EC @ 2 ml/lit when situation warrants.

Epilachna beetle

- Collect and destroy severely affected leaves along with grubs, pupae and beetles.
- Spray fipronil 2 ml/lit.

Whitefly

- Monitor the incidence using yellow sticky trap @ 12/ha.
- Spray Neem oil 3 ml/lit + Teepol 1 ml/lit or NSKE 5% (50g /lit). 223

8.4.2 Pest Management In Tomato

Fruit borers

- Plant 40 days old marigold (American Tall) as trap crop with 25 days old tomato seedlings @ 1:16 row ratio.
- Set up pheromone traps @ 12/ha.
- Collect and destroy infested fruits, leaves, egg and gregarious larvae.
- Based on ETL (5% fruit damage) spray quinalphos 2.5 ml/lit (or) Bacillus thuringiensis @ 2g/lit.
- Release twice, Trichogramma chilonis @ 50,000/ha release from flowering onwards at 10 days interval.

- Spray HaNPV (or) SlNPV @ 1.5 x 1012 POSB/ha in the evening hours.

- For Spodoptera litura, poison baiting with carbaryl 50 WP - 1.25 kg, rice bran 12.5 kg, jaggery 1.25 kg and water 7.5 lit per hectare. vii Grow resistant varieties like T27, T32.

Serpentine leaf miner

Spray NSKE 5%.

8.4.3 Pest Management in Bhendi

Sucking pests (Leaf hopper, aphids and whiteflies)

- Grow whitefly tolerant varieties like Arka Anamica, Hisar Unnat, Varsha Uphar or P7 (or) fruit borer resistant varieties like Parkins Long Green, Karnal special.

- Spray dimethoate 30 EC 2 ml/lit, or Neem Seed Kernel Extract 5% (50 g/lit). 224

- **Fruit borers**

- Setting up pheromone traps @ 12/ha.

- Collection and disposal of infested plant parts.

- Release Trichogramma egg parasitoid @ 1.0 lakh/ha.

- Release first instar grubs of Chrysoperla carnea @ 10,000/ha.

- Spray Bacillus thuringiensis @ 2 g/lit.

- Based on ETL (5% fruit damage) spraying cartap hydrochloride 5.0 sp 2g/lit or cartap hydrochloride 5.0 wp 1g/lit. combined with NSKE 5%.

Mites

Spraying either wettable Sulphur 50 WP 2 g/lit or dicofol 3 ml/lit.

Nematodes

In endemic areas, apply carbofuran 3G @ 33 kg/ha (or) phorate 10G @ 10 kg/ha with Neem cake @ 400 kg/ha at sowing in furrows along with fertilizers.

8.4.5 Pest Management in Cucurbits

Pumpkin beetles and leaf caterpillars

- Early planting of pumpkin during October – November

- Frequent raking of soil beneath the crop to expose and kill the eggs and grubs.

- Hand collection and destruction of infested leaves and fruits.

- Spray malathion 50 EC 1 ml/lit, dimethoate 30 EC 2 ml/lit, methyl demeton 25 EC or fenthion 100 EC 1 ml/lit. 225

Fruitfly

- In endemic areas, sowing time may be adjusted in such a way that fruiting should not coincide with monsoon.
- Fruit fly resistant pumpkin varieties like Arka Swarramuki may be grown.
- Ribbed gourd may be grown as a trap crop and carbaryl 50 WP 2
- g/lit (or) malathion 2 ml/lit, may be sprayed on the congregating adult flies on the under surface of leaves.
- Attractants like citronella oil, eucalyptus oil, acetic acid (vinegar) dextrose and lactic acid may be used to trap adult flies.
- Poison baiting may be employed with saturated sugar solution 5 ml
- + malathion 50 EC 5 ml + 100 ml fermented palm juice. This mixture may be kept in earthern vessels in many places in the field.
- Use fishmeal trap to attract and kill the flies. Take 5 g of wet fishmeal in a (20 x 15 cm) polythene bag. Make six holes (3 mm dia.) around the periphery of the bag at equidistance at about 2 cm from the bottom of the bag. Impregnate an absorbent cotton plug with 1 ml of dichlorvos and keep this also inside the poly bag. Suspend such fish meal traps at places in the field @ 50/ha. Dichlorvos should be replenished every week and fishmeal has to be replaced once in 20 days. 226

Root-knot nematode

In endemic areas, apply carbofuran 3 G @ 33 kg/ha with Neem cake @ 400 kg/ha before sowing.

Leaf miner

Spray Neem Seed Kernel Extract 5%.

Caution

In cucurbits, DDT, Lindane 1.3 D, Copper Oxychloride, Bordeaux mixture and Sulphur dust should not be used as these are highly phytotoxic.

8.4.6 Pest Management In Cruciferous Vegetables

Cutworms

- Setting up light traps during summer months to attract the moths.
- Install srinkler system of irrigation and irrigate during day time to expose the larvae to bird predation.
- Drench the collar region of the plants with chlorpyriphos 2 ml/lit (or) one day after planting (or) apply lindane 1.3 D @ 10 kg/ha in soil before planting.

- Collect and destroy the weed i Gynandropis pentaphylla. 227

White grub

- Summer ploughing.

- Dusting or quinalphos 5 D @ 25 kg/ha, 10 days after summer rains.

- Operate light traps between 7 and 9 PM during April - May to attract adults.

- Pre-sowing soil application of entomopathogenous fungus Metarrhizium anisopliae @ 20 kg/ha and rake the soil during May.

- In endemic areas, apply phorate 10 G @ 25 kg/ha during August - October.

Cabbage aphid

- Monitor and attract the aphids using yellow sticky trap @ 12/ha.

- Spray Neem Oil 2% (20 ml/lit) (or) dimethoate 30 EC @ 2 ml/lit along with Teepol @ 0.5 ml/lit.

Diamondback moth

- Grow two rows of mustard as trap crop at the end of every 25 rows of cabbage. The first row of mustard should be sown 15 days prior to planting of cabbage or 20 day old mustard seedlings should be planted along with cabbage. The second row of mustard should be sown 25 days after planting of cabbage. The mustard crop should be periodically sprayed with Dihlorvos @ 1 ml/lit once in 10 days to check the migration of Diamondback moth.

- Install pheromone traps to attract and monitor the moths @ 12/ha. 228

- Based on the ETL (2 larvae/plant), spray cartap hydrochloride 1 g/lit, Baillus thuringiensis 2 g/lit or NSKE 5% (50 g/lit) at primordial stage (ca. 17-25 DAP). Spray fluid should be combined with sticking agents like Teepol or Sandovit (0.5 ml/lit).

- Release larval parasitoids @ 20,000/release starting from 20 DAP at fortnightly interval. Five such releases are effective against diamondback moth. In plains release Cotesia plutellae (Braconidae: Hymenoptera) and in hills Diadegma semiclausum (Ichneumonidae: Hymenoptera).

8.4.7 Pest Management in Moringa

Fruit fly

- Raking the soil beneath the crop canopy after application of NSKE 5% or Lindane 1.3 D @ 25 kg/ha.

- Collect and destroy the fruits which ooze out or rotten.

- Based on the ETL (15% infested pods), spray dichlorvos 1 ml/lit (or) fenthion 1.5 ml/lit when the pods are 20-30 days old.

Budworm, leaf caterpillar, leaf webber

Spraying dichlorvos 1 ml/lit or dusting carbaryl 10 D @ 25 kg/ha.

Hairy caterpillars

Burning the congregating caterpillars on the bark with flame thrower / burning flame. 229

8.4.8 Pest Management in Potato

Cut worms

- Deep ploughing to expose the pupae and larvae to predators.
- Irrigate with sprinklers during day time to expose the larvae to predation.
- Drench the collar region of the plants in evening hours with chlorpyriphos or endosulfan 2 ml/lit one day after planting.

Aphid, Leaf hopper

Spray methyl demeton 25 EC (or) dimethoate 30 EC @ 2 ml/lit (or) acephate 75 SP 1 g/lit.

White grub

- Summer ploughing.
- Dust (or) quinalphos 5 D @ 25 kg/ha, 10 days after first summer rains.
- Operate light traps between 7.00 - 9.00 PM during April - May.
- Soil application of entomopathogenous fungus, Metarrhizium anisopliae @ 20 kg/ha and rake the soil during May.
- Hand picking adult beetles in the morning hours.
- In endemic areas, apply Phorate @ 10 G at 25 kg/ha in soil during August - October. 230

Potato tuber moth

- Deep planting of tubers at 10-15 cm depth.
- Pheromone traps @ 20/ha both in field and godowns.
- Earthing up at 60 DAP, to avoid oviposition by moths.
- When the foliar damage exceeds the ETL (5%), spray NSKE 5% or quinalphos 20 EC 2 ml/lit.
- In godowns, the upper surface of potato leaves should be covered with either Lantana (or) Eupatorium leaves as oviposition deterrents against moths.
- Seed tubes may be treated with quinalphos or dust @ 1 kg/ 100 kg of tubers.

8.4.9 Pest Management in Sweet Potato

Sweet potato weevil

- Maintaining field sanitation by removing crop residues, debris and alternative hosts.
- Selecting weevil-free planting materials.
- Sweet potato tubers are sliced into pieces of about 100 g and placed 5 m apart in the field from 4.00 PM to attract the weevils. Next day morning, the tubers should be collected with attracted weevils and meticulously disposed.
- Dip the planting material in fenthion 100 EC, fenitrothion 25 EC before planting.
- Rake up the soil and earth up 50 DAP.
- Drenching the soil with or fenthion 100 EC @ 2 ml/lit. These insecticides may also be sprayed when necessary. 231
- Harvesting the tubers immediately after maturity and destroying residues.
- Installing yellow sticky traps @ 12/ha.
- In storage, the tubers may be covered with sand.

Caterpillars, tortoise beetles

- Collecting and destroying larvae, damaged leaves and beetles.
- Spraying Fipronil 2 Ml/Lit.

8.4.10 Pest Management in Topioca

Scales

- Careful selection of setts free from scale insects.
- Stacking the setts in shade in vertical position.
- Dipping the setts in dimethoate 0.03% or methyl demeton 0.025% for 10 minutes before planting.
- Encourage the predatory ladybird beetle, hiloorus nigritus.

Whitefly

- Maintaining field hygiene by removing alternative weed hosts like Abutilon indicum.
- Installing yellow sticky traps @ 12/ha.
- Avoid excess irrigation and nitrogen.
- Spray Neem Oil 5 ml/lit (or) Fish Oil Rosin Soap 20 g/lit (or) methyl demeton 25 EC 2 ml/lit (or) Phosalone 35 EC 2 ml/lit. Avoid synthetic pyrethroids. 232

8.4.11 Pest Management in Turmeric

Pre planting treatment : The turmeric rhizomes should be dipped in a mixture of carbendazin 50 WP 1 g/lit + phosalone 35 EC 2 g/lit (or) monocrotophos 36 WSC 1.5 ml/lit.

Rhizomone Scale : Well rotten sheep manure / poultry manure should be applied in two splits @ 10 tons/ha, first before planting and the second at the time of earthing up.

8.4.12 Pest Management in Chillies

Thrips : Spray NSKE 5%, dimethoate 30 EC 2 ml/lit, methyl demeton 25EC 2 ml/lit, formothion 2 ml/lit, quinalphos 1.5 D @ 20 kg/ha thrice at fortnightly intervals.

Aphids : Spray acephate 75 SP 1 g/lit, methyl demeton 25 EC 2 ml/lit or phosalone 35 EC 2 ml/lit.

Yellow Muranai Mite (Broad Mite) : Spray dicofol 18.5 EC 3 ml/lit, ethion 50 EC 4 ml/lit or wettable sulphur 50 WP @ 6 g/lit.

Fruit borers :

- ⦿ Setting up pheromone traps for Helicoverpa armigera (or) Spodoptera litura @ 12/ha.

- ⦿ Collection and destruction of grown up caterpillars and damaged fruits.

- ⦿ Poison baiting with rice bran 5 kg, jaggery 500 g, carbaryl 50 WP 500 g and water 3 lit per acre in the evening hours.

- ⦿ Spray chlorpyriphos 20 EC 3 ml/lit or quinalphos 25 EC 2.5 ml/lit. 233

8.4.13 Pest Management in Cardamom

Thrips :

- ⦿ Regulating shade in such a way to have partial shade.

- ⦿ Spray or phosalone 35 EC @ 1 ml/lit.

Shoot and Fruit borer

- ⦿ Collection and destruction of infested plant parts before spraying.

- ⦿ Spraying monocrotophos 36 WSC @ 2.5 ml/lit or phosalone 35 EC @ 3 ml/lit.

Hairy caterpillars

Spray phosalone 35 EC @ 1 ml/lit.

Rhizome weevil

Drench lindane 20 EC @ 2 ml/lit.

Mites

Spray dicofol 18 EC @ 2 ml/lit

Aphids (Katte disease vector) : Spraying regularly methyl demeton, 25 EC, dimethoate 30 EC.

Red flour beetle : Storing capsules in alkathene lined jute bags sprayed with malathion 0.1%. Fumigation with methyl bromide.

8.4.14 Pest Management in Pepper

Leaf gall thrips

    i.   Raking the soil and applying quinalphos 1.5 D @ 20 kg/ha.

    ii. Spray anyone of the following insecticides three rounds at monthly intervals starting from new flush formation. 234

Dimethoate 30 EC @ 2 ml/lit

Chlorpyriphos 20 EC @ 2 ml/lit

Dhchlorvos 76 WSC @ 1 ml/lit

Pollu beetle

- Spray fipronil 2 ml/lit once in July and October

Scales

- Removal of severely affected plant parts
- Spray methyl demeton 2 ml/lit or dimethoate 2 ml/lit

8.4.15 Pest Management in Betelvine

Scale insect, mealy bugs, aphids :

- Selection of infestation free vines for planting.
- Spraying chlorpyriphos 20 EC @ 2 ml/lit, malathion 50 EC @
- 1 ml/lit, dimethoate 30 E @ 2 ml/lit NSKE 5% along with teepol 0.5 ml/lit.

Root-knot nematode

- Apply Neem Cake @ 1 ton/ha shade-dried Calotropis leaves @ 2.5 tons/ha to soil after lowering vines.

Mites

Spray wettable sulphur 50 WP @ 1 g/lit or dicofol 18 EC @ 0.5 ml/lit.

Caution

- Insecticides should be applied only after harvesting leaves
- After sprayings a waiting period of 3 weeks should be strictly observed 235

8.4.16 Pest Management in Mango

Infloresence hoppers, shoot webber

- Spray two rounds of acephate 75 SP @ 1 g/lit, phosalone 35 EC @ 1.5ml/lit, or phosphamidon 85 WSC @ 1 ml/lit. First at the time of panicle emergence and the second a fortnight later.

- Phosphamidon 85 WSC @ 1 ml + Neem Oil 5 ml/lit may be sprayed against both hoppers and shoot webber

Leaf galls and aphids

Spray dimethoate 30 E or methyl demeton @ 2 ml/lit.

Flower webber

Spray phosalone 35 EC @ 2 ml/lit.

    Net weevil : Spray fenthion 100 EC @ 1 ml/lit twice. First at the marble stage and the second a fortnight later.

Stem borer

- Avoiding injuries at the base of trunk while pruning and removing alternative hosts like moringa in the near vicinity.

- During off-season, padding with 10 ml monocrotophos soaked in absorbent cotton per tree without unnecessarily injuring the trunk.

- Using a needle or long wire, the grubs may be hooked out through the bore holes. The bore holes may be filled with carbofuran 3 G @ 5 g/tree and plugged with clay + fytolon paste. 236

Fruit fly

- Interspaces may be ploughed to expose and kill the soil borne puparia.

- The infested and fallen fruits should be carefully disposed of.

- Apply a bait-spray combining anyone of following insecticides with molasses or jaggery (10 g/lit) two rounds at weekly interval before ripening

Fenthion 100 EC @ 1 ml/lit (or)

Malathion 50 EC @ 2 ml/lit.

8.4.17 Pest Management in Citrus Fruits
Leaf miner

- Spray NSKE 5% (50 g/lit), Neem Cake Extract 5% or Neem Oil 3 ml/lit.

- Spray dichlorvos 76 WSC 1 ml/lit, dimethoate 2 ml/lit or fenthion 1 ml/lit.

Leaf caterpillar

- Hand picking and destroying the greenish brown larvae
- Spray endosulfan 2 ml/lit.

White fly, black fly and aphids

- Spray quinalphos 2 ml/lit, methyl demeton 1 ml/lit, Neem Oil 3% or Fish Oil Rosin Soap 30 g/lit.

Rust mite

- Spray dicofol 18 EC 2.5 ml/lit, or wettable sulphur 50 WP 2 g/lit.

Fruit sucking moths

- Destroy the weed host, Tinospora cordifolia
- Apply smoke and set up light traps wherever possible
- Set up food lures with rotten tomatoes (or) pieces of citrus fruits
- Cover the fruit with perforated poly bags
- Bait with fermented molasses / jaggery + malathion 1 ml/lit

Fruit fly

- Interspaces may be ploughed to expose and kill the soil borne puparia.
- The infested and fallen fruits should be carefully disposed of.
- Apply a bait-spray combining anyone of following insecticides with molasses or jaggery (10 g/lit) two rounds at weekly interval before ripening

Fenthion 100 EC @ 1 ml/lit (or)

Malathion 50 EC @ 2 ml/lit. 237

Nematodes

- Apply Pseudomonas florescens formulation @ 20 g/tree at
- 15 cm depth 50 cm away from the trunk once in four months.

Stem borer

- Prune the branches infested
- Plug the fresh boreholes with absorbent cotton soaked in monocrotophos 5 ml/20 ml water

8.4.18 Pest Management in Sapota

Leaf webber : Spray phosalone 35 EC 2 ml/lit.

Hairy caterpillars : Spray chlorpyriphos 25 EC or phosalone 35 EC 2 ml/lit.

Bud worm : Spray phosalone 2 ml/lit, phosphamidon 1 ml/lit, or NSKE 5%.

8.4.19 Pest Management in Guava

Tea mosquito bug : Spray anyone of the following in the early morning hours or late evening hours at 21 day intervals four times minimum. Neem Oil 3%, malathion 1 ml/lit, fenthion 1 ml/lit or endosulfan 2 ml/lit.

Mealy bug :

- Release ryptolaemus montrouzieri beetles @ 10/tree
- Spray Neem Oil 5 ml + Triazophos 2 ml/lit or Neem Oil 5 ml + Phosalone 2 ml/lit. 239 Fruit fly i. Disposal of infested fruits
- Raking the soil and flooding for 24 h
- Make annihilation technique using methyl eugenol 0.1 ml + dichlorvos -0.04% in cotton wool as in mango.
- Spray malathion or fenitrothion 1 ml/lit
- Drench soil with NSKE 5%

8.4.20 Pest Management in Pomegranate

Anar butterfly (or) Fruit borer

- Bagging the fruits with polythene covers
- Spray NO 3% or NSKE 5% twice when insect activity is noticed
- Release egg parasitoid Trichogramma chilonis @ 1 lakh/acre
- Spray dimethoate 1.5 ml/lit based on the ETL of 5 eggs/plant

8.4.21 Pest Management in Banana

Rhizome (or) Corm weevil

- Trapping the adult weevils by placing chopped pseudostem in the cropped area
- Selecting infestation - free suckers
- Soil incorporation of lindane 1.3 D 20 g/plant, 10-20 g/plant arbofuran 3G 10 g/plant or phorate 10 G 5 g/plant around pseudostem. 240

Pseudostem weevil

- Disposal of infested trees by chopping and burning
- Maintaining healthy plantation by periodical removal of dry leaves and suckers
- Pseudostem injection with monocrotophos (50 ml + 350 ml water0 @ 2ml at 45 cm height and another @ 2 ml at 150 cm height from ground level at monthly intervals from 5th - 8 th months. Beyond 8 months (after flowering), this should not be done.

Banana aphid (Vector of Bunchy top disease)

- ⊙ Spray methyl demeton 2 ml/lit, phosphamidon 1 ml/lit, midechlopride 0.5 ml or dimethoate @ 2 ml/lit towards the crown and pseudostem base thrice at 21 day intervals.

- ⊙ Pseudostem injection of monorotophos 1 ml in 4 ml of water per tree at 45 day interval from the 3 rd month till flowering using "TNAU - Banana Injector".

- ⊙ Avoid monocrotophos after flowering. Thrips, Lacewing bug : Spray methyl demeton 2 ml/lit, monocrotophos @ 1 ml/lit or phosphamidon @ 1 ml/lit.

8.4.22 Pest Management in Cashew

Stem and Root Borer

- ⊙ Periodical cleaning of collar region, removal of grubs, pupae and eggs and inter ploughing wherever possible during monsoon months. 241

- ⊙ Swabbing the bark of the exposed roots and shoots with carbaryl 50 WP 2 g/lit, lindane 20 EC 1 ml/lit or coal tar +

- ⊙ Kerosene - Coaltar mixture (1:2) upto one metre height on the trunk and on exposed bark after shaving the infested bark.

- ⊙ Root-feeding with monocrotophos 10 ml + 10 ml water in a small polythene bag twice a year on both sides of the trunk.

Tea mosquito bug

- ⊙ Regulating the shade to facilitate proper penetration of sunlight inside the canopy.

- ⊙ Spray the following insecticides, thoroughly covering foliage and bark during early morning hours. Monocrotophos @ 2 ml/lit at new flush formation during November - December. Malathion / Chlorpyrites 2 m g/lit + Urea 3% at flower initiation during January - February and again at fruiting time during March - April.

Note: Monocrotophos should not be sprayed at flowering time.

8.4.23 Pest Management In Grapevine

Flea beetle

- ⊙ Loose bark may be removed at the time of pruning.

- ⊙ Spray phosalone @ 2ml/lit, quinalphos @ 2 ml/lit or immediately after pruning and repeated 2-3 times. 242

2. Thrips

Spray methyl demeton or dimethoate 2 ml/lit

3. Mealy bug

- Release coccinellid beetle Cryptolaemus montrouzieri @ 10 / vineii. Apply quinalphos or methyl parathion dust in soil @ 20 kg/ha to kill phoretic ants
- Spray methyl demeton or monocrotophos 2 ml/lit iv. Spray dichlorvos @ 1ml / lit + Fish Oil Rosin Soap @ 25 g/lit
- Stem girdler Swabbing the trunk with carbaryl 50 WP 2 g/lit

Note: Waiting period for dimethoate and carbaryl is five days.

Fruit bat

Covering with nets and smoking in the evening hours.

8.4.25 Pest Management in Coconut

Rhinoceros beetle

- Destroy and dispose of all dead trees
- Avoid manure pits in the vicinity of coconut gardens
- Rake and turn up the decaying manure to expose the developing grub, egg and pupae to sun drying and predation. Then apply the fungal culture of Metarrhizium anisopliae to manure pits during cooler months of October - December. 243
- Encourage reduviid predators, Platymeris laevicollis
- Once in three months, drench the manure pits with Carbaryl 50 WP 1 g/lit
- In seedlings, place naphthalene balls @ 3 / tree, in the innermost three leaf axils once in 45 days.
- Soak castor cake @ 1 kg/5 lit of water in wide mouthed mud pots and keep them in the garden to attract and kill adults. Replace the slurry once in 30 days.
- Fermented toddy may be kept in wide mouthed earthern vessels in different places to attract the adults during night.
- The crown region may be properly cleaned during harvests and the adults may be hooked out using a long wire.
- Light traps may be set up to attract the adults during monsoon months and following rains during summer.
- The top-most three axils may be filled with a mixture of Sand + Neem Seed Powder (2:1) once in three months (150 g/tree)
- Use Aggregation pheromone traps.

Red palm weevil

- Removal and disposal of damaged and wilted trees.

- Avoiding injuries on trunk. Any injury should be plastered with clay or cement with Fytolon.
- Avoid cutting green fronds.
- Avoiding incidence of Rhinoceros beetle. 244 v. Root feeding with monocrotophos @ 10 ml + 10 ml water after harvesting nuts again, only after 45 days nut should be harvested.
- Setting up attractant traps using mud pots with molasses / toddy 2.5 lit + acetic acid 5 ml + yeast 5 g + split tender coconut stems / petioles @ 30/ac.
- Insert 1-2 Aluminium phosphide tablets inside the tunnel and plug all the holes with clay + Fytolon.
- Use aggregation pheromone traps.

Black headed caterpillar

- Cutting and burning all the infested leaves and fronds.
- In small plantations, carbaryl 50 WP 2 g/lit may be sprayed.
- In summer months release Bethylids and Braconid and Eulophid parasitoids from January onwards at 1:1:10 per tree.
- Root feeding with monocrotophos @ 10 ml + 10 ml water with a waiting period of 45 days after root feeding.

Shot - hole borer, bark weevil

- Swab the stem with Carbaryl 50 WP 2 g/lit.
- Root feeding with monocrotophos 10 ml + 10 ml water

Eriophyid mite - IPM package

- **Nutrients (per tree / year)**

Urea 1.3 kg, Super 2.0 kg, Potash 3.5 kg, Neem cake 5 kg, Borax 50 g, Gypsum 1 kg, $MgSO_4$ 500 g, FYM 50 kg 245

- **Root feeding**
  a. Root feeding with TNAU - Agro Biocide 30 ml/tree
  b. Root feeding Carbosulfan 15 ml + 15 ml water / tree (45 days after)
  c. TNAU - Agro biocide - 30 ml/tree - (60 days after Carbosulfan root feeding).

Note: Before root feeding, pluck nuts. After root feeding, next harvest should be done 45 days later.s

Rodents

- Fixing inverted cone shaped tin sheets on stem

- Wrapping tin sheets to a length of 1-2 feet on stem 5 m above ground level (or with any thorny plant materials like Prosophis)

- Placing Bromodiolone 0.005% @ 10 g/tree at crown region at regular intervals.

8.4.26 Pest Management in Coffee

White borer :

- Arabica coffee grown under inadequate shade is highly prone to the attack. Provide optimum shade.

- Trace the infested plants prior to the adult flight periods (March and September) by tracing the ridges on the stem. Avoid injuries on stem and roots.

- Uprooted stem / plants should not be heaped inside the plantations.

- Remove the loose scaly bark of main stem and primaries using coir glove or coconut husk to remove cracks and crevices on which eggs are normally deposited. Do not use any sharp implements. 246

- Spray and swab the main stem and thick primaries once in April - May and October - December with Lindane 20 EC 1.25 lit + 200 ml Teepol in 200 lit water at the time of peak adult activity (March and September) NSKE 5% also can be employed in more frequencies.

Shot hole borer

- Robusta coffee is more prone to the attack under heavy shade.

- Pruning branches and spray endosulfan @ 2 ml/lit.

Green scales and Mealy bugs

- Spray Verticillum lecanii fungus @ 6 x 106 spores/ml

- Release Cryptolaemus montrouzieri @ 300/ac.

- Spray quinalphos 2 ml/lit, fenthion 1 ml/lit or fenitrothion 1 ml/lit

Berry borer

- Maintaining thin shade and proper training of the plant.

- Harvesting should be perfect without any left over beans on plants soil.

- The left over harvest (gleaning) reduces the inoculum to a great extent.

- Drying the berries to the following moisture levels. Parchment 10%; Arabica cherry 10.5%; Robusta cherry 11.0%. v. Spray Beauveria bassiana fungus (white muscardine fungus).

- Seed beans may be transported after thorough disinfestations. 247

8.4.27 Pest Management in Tea

Scales : Spraycarbaryl 2 g/lit, endosulfan 2 ml/lit, quinalphos 2 ml/lit or chlorpyriphos 2 ml/lit.

Sahyadrassus borer :

⊙ Clean the base of bush

⊙ Kill the hiding larvae by inserting a thick wire into the borer hole.

⊙ Inject quinalphos 2 ml using a syringe or ink filler through the borerhole and plug with moist clay.

Mites : Spray dicofol 2 ml/lit, sulphur 40% 2 g/lit, sulphur 80% 1 g/lit or ethion 50 EC 1ml /lit.

8.4.28 Pest Management in Jasmine

Budworm / blossom midge : Spray monocrotophos 2 ml/lit or endosulfan 2 ml/lit

Red spider mite and Erineum mites : Sulphur 50 WP 2 g/lit or dicofol 2.5 ml/lit.

8.4.29 Pest Management in Rose

Beetles : Hand picking the beetles during dry time and spray endosulfan 2 ml/lit.

Scale :

i. Severely infested branches should be cut and burnt.
ii. Spray endosulfan 2 ml/lit, malathion 2 ml/lit FORS 25 g/lit or carbofuran 3 G 5 g/plant at the time of pruning and during March - April. 248

Mealy bug : Spray monocrotophos 2 ml/lit or methyl parathion 2 ml/lit.

Bud worm : Spray monocrotophos 2 ml/lit.

Thrips, aphids and leaf hoppers : Spray methyl demeton 2 ml/lit (or) NO 3% (or) apply carbofuran 3G 10 G/Plant.

8.4.30 Pest Management in Subabul

Psyllid

i. Spray a strong jet of water on new flushes.
ii. Spray acephate 1 g/lit or triazophos 1 ml/lit and following a waiting period of 7 days before leaf harvest.

9

GREENHOUSE INSECT MANAGEMENT

9.1 INTRODUCTION

The warm, humid conditions and abundant food in a greenhouse provide an excellent, stable environment for pest development. Often, the natural enemies that serve to keep pests under control outside are not present in the greenhouse. For these reasons, pest situations often develop in this indoor environment more rapidly and with greater severity than outdoors. Pest problems can be chronic unless recognized and corrected. Successful control of insect pests on greenhouse vegetables and ornamentals depends on several factors. Proper cultural practices can minimize the chance for initiation and buildup of infestations. Early detection and diagnosis are keys to greenhouse pest management, as well as the proper choice and application of pesticides when they are needed. The pests that attack plants produced under conventional greenhouse practices also infest plants produced in float systems. Float systems are especially prone to problems with fungus gnats, shore flies and bloodworms. Some greenhouse insects can transmit diseases to the plants which are often more serious than the feeding injury that the insect causes. These insect "vectors" include some aphids, leafhoppers, thrips and whiteflies. In these instances, the diseases must be managed through early insect control.

9.2 GREENHOUSE PESTICIDE SAFETY

While pesticides are important tools used in managing greenhouse pests, their use in enclosed spaces increases the potential for worker exposure during and after application. PAT-4, Greenhouse Pesticides and Pesticide Safety, contains information on general precautions needed when using pesticides in the greenhouse, appropriate safety equipment, calibration and application, as well as specific information on pesticides registered for greenhouse use. PAT-4 is available at your county Cooperative Extension Service office.

9.3 COMMON GREENHOUSE INSECTS AND RELATED PESTS

Since greenhouse conditions allow rapid development of pest populations, early detection and diagnosis of pest insects are necessary to make control decisions before the problem gets out of hand and you suffer economic loss. Some common and important greenhouse pests to keep a close watch for are aphids, fungus gnats, thrips, whiteflies, caterpillars, leafminers, mealybugs, mites, slugs and snails.

9.3.1 Aphids

Aphids or plant lice are small, soft-bodied, sluggish insects that cluster in colonies on the leaves and stems of the host plants. They are sucking insects that insert their beaks into a leaf or stem to extract plant sap. They are usually found on and under the youngest leaves, and, in general, prefer to feed on tender, young growth.

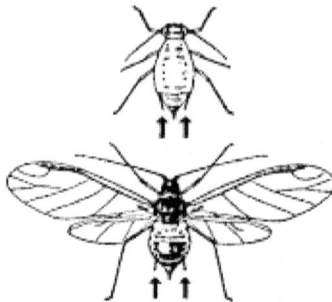

Fig. 1: Winged and wingless aphids

Aphids are the only insects that have a pair of cornicles, or tubes that resemble exhaust pipes, on their abdomen. Aphids multiply rapidly. In greenhouses, each one is a female capable of giving live birth to daughters in about seven days after its own birth. These asexually reproducing female aphids may be winged or wingless. Adult aphids can give birth to six to ten young per day over their 20- to 30-day life span. Enormous populations can build up in a relatively short period. Feeding by aphids can cause leaves or stems to curl or pucker; this leaf distortion often protects the aphids from contact insecticides. Much of the sap they suck from the plant passes through their bodies and is dropped on the leaves as "honeydew." Ants, which feed on honeydew, are often found in association with aphid infestations. Black sooty mold often develops on leaves with honeydew. Aphids can also transmit serious viral diseases. Managing these diseases usually requires control of the insect that transmits the disease. Aphid infestations usually begin with winged individuals entering the greenhouse through openings. Insecticide applications to control aphids often must be repeated to manage infestations. Usually, two to three applications spaced at three- to seven-day intervals, depending on the severity of an infestation, are necessary. Insecticide products need to be alternated for aphid control to delay development of resistance.

If you observe aphids that appear tan or off-color relative to the other aphids, they may be parasitized aphids known as "mummies." These naturally-occurring wasp parasites so important to aphid control are smaller than aphids. When these parasites emerge, they cut a round hole in the upper portion of the abdomen of the dead aphid and begin to search for their prey.

9.3.2 Fungus Gnats, Shore Flies and Bloodworms

The high humidity and moist organic growing media in greenhouses provide an excellent breeding area for several types of gnats. These insects are abundant outdoors where they can breed in virtually any accumulation of standing water that remains in place for several days.

Fungus Gnats: Fungus gnat larvae can be serious pests of some greenhouse plants. The larvae of most species are scavengers, feeding on decaying organic matter in the soil. However, larvae of some species will feed on root hairs, enter the roots or even attack the crown or stem of the plant. Plants infested with fungus gnats generally lack vigor and may begin to wilt. Adults are frequently observed running on the foliage or medium before injury caused by the larvae becomes apparent.

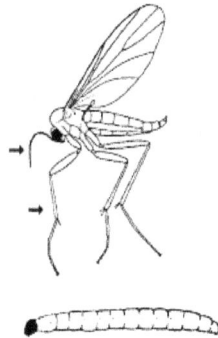

Fig. 2: Fungus gnat and larva

Fungus gnats are small (1/8 inch) black flies with comparatively long legs and antennae, tiny heads and one pair of clear wings. Females lay tiny ribbons of yellowish-white eggs in growing media that hatch within four days. The clear larvae are legless and have black heads.

Larvae mature underground in about 14 days and pupate near the surface of the medium. They construct a pupal case made of soil debris. Adults live only about a week. Under greenhouse conditions, about 20-25 days are required to complete a generation. Larvae are somewhat gregarious and are found in clusters in the soil.

Shore Flies Shore flies are gnat-like insects similar to fungus gnats. They differ in having short antennae, red eyes and heavier dark bodies. A pair of smoky wings with several clear spots can be seen when looking closely at the insect. They are good fliers and can be seen resting on almost any surface in the greenhouse. They resemble winged aphids, but aphids have two pairs of wings and the distinctive, tube-like cornicles on the abdomen.

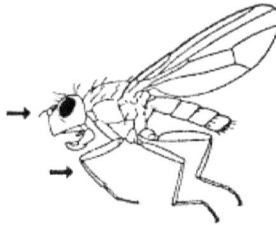

Fig. 3: Shore fly

Their life cycle is similar to that of the fungus gnat. The yellow to brown larvae, which may be up to 1/4-inch long, differ in having no apparent head. Both larvae and adults feed mostly on algae growing on media, floors, benches or pots. They rarely damage plant tissue, but the mobile adults may spread soil pathogens inside the greenhouse.

9.3.4 Bloodworms

Bloodworms are the striking red "worms" that may be seen wriggling in float plant water. These long, cylindrical larvae are similar to fungus gnat larvae in lacking legs and having a distinct brown head. The red is due to the presence of hemoglobin, the same oxygen-carrying material present in human blood. The presence of hemoglobin allows this insect to develop in water with a very low oxygen content.

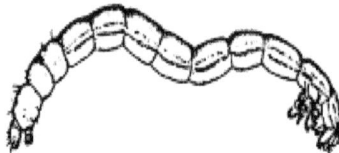

Fig. 4: Bloadworm

Bloodworms are common in stagnant water, animal watering troughs and other accumulations of standing water. These insects are close relatives of the mosquito, but the adults do not have sucking mouthparts and are not blood feeders. The larvae have chewing mouthparts and generally feed on algae or other organic matter in the water. They may be found in plant roots that grow through the bottoms of float trays but apparently do not cause significant injury.

While fungus gnats and shore flies live in "very wet" situations, bloodworms generally live entirely in water. Eliminating standing puddles around the area and keeping to a minimum the amount of exposed water surface in the float bed will reduce the presence of these insects.

Avoiding excessive watering to reduce moisture in growing media will help regulate these pests because they require high moisture. Highly organic soils and potting mixtures containing peat are attractive to egg-laying fungus gnats. Sprays or drenches containing Bacillius thuringiensis Serotype H-14 (Gnatrol) can be used to control fungus gnat larvae on ornamentals and nursery plantings in the greenhouse. This treatment is not effective against shore flies.

9.3.5 Thrips

Thrips are tiny, slender insects about 1/25-inch long. They range in color from light brown to black. They have four wings, each fringed with a row of long hairs, that are held flat over their back. Plant-feeding thrips cause economic damage when they infest the flowers, buds and young fruits of a crop.

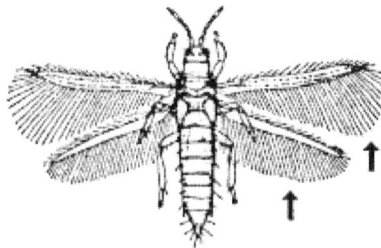

Fig. 5: Winged thrips

Thrips feed by rasping the plant surface and sucking up the exuding sap. Heavily infested leaves have a mottled or silvery appearance. Female thrips insert eggs into slits in the leaf. Eggs hatch in two to seven days. Nymphs feed much like adults and molt four times during development. They are inactive during the last nymphal stage before becoming an adult.

Winged adults are carried into the greenhouse on contaminated plant material, or they fly in during the summer and continue to breed throughout the winter. Preventing infestations through the use of screens on ventilators, inspecting new material entering the greenhouse and controlling weeds in the greenhouse will help to manage thrips.

Several species occur in greenhouses. Thrips attack a wide range of plants in the greenhouse. Highly susceptible hosts include azalea, calla lily, croton, cyclamen, cucumber, fuchsia, ivy and rose. Various thrip species also transmit plant diseases. The most serious are the western flower thrips and onion thrips, which are vectors of tomato spotted wilt virus or impatiens necrotic spot virus. This virus attacks a wide variety of plants.

9.3.6 Greenhouse and Sweet Potato Whiteflies

Whiteflies are serious pests in the greenhouse and are often seen on fuchsias, poinsettias, cucumbers, lettuce and tomatoes. Through regular monitoring, these preferred hosts can be used as indicator "plants," alerting greenhouse managers to the first signs of whitefly infestations. These powdery white insects, about 1/12 inch in length, flutter from the undersides of leaves when the plants are disturbed. The lower surface of the leaves may be infested with all life stages of whiteflies.

The female of these sap-sucking insects may lay 150 eggs at the rate of 25 per day. The newly emerged crawler moves only a short distance before settling down to feed. After three larval molts, the pupal stage is formed, from which the adult emerges. The entire life cycle takes 21-36 days, depending on the greenhouse environment.

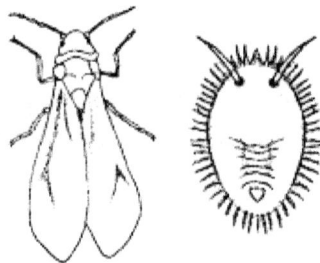

Fig. 6: Whitefly and larva

The greenhouse and sweet potato whiteflies are similar in appearance but differ in their biology and control. Both whitefly species develop entirely on the undersides of leaves. Their life cycle may be as short as 20 to 25 days.

The sweet potato whitefly has a broader host range, higher reproductive potential, stronger resistance to insecticides and a powerfully phytotoxic enzyme system. This whitefly is a vector of gemini viruses in tomatoes. Control of these viruses relies on proper sanitation and control of the whitefly vectors.

Insecticides used to control adult whiteflies are usually ineffective against immatures. Because adult whiteflies often continue to emerge after these applications, insecticides used to control adults must be applied frequently, two to three times with three- to four-day intervals between sprays, to control infestations. Growth regulators used to control immature stages can be applied less frequently, at seven- to 14-day intervals as necessary, to control infestations.

A tiny parasitic wasp, Encarsia formosa, attacks the larval stage of whiteflies and sometimes occurs naturally in greenhouses. While they are not useful in controlling heavy whitefly infestations, they can be used successfully against early infestations under conditions that favor their development over the development of whiteflies (64°- 80°F).

After the parasitized larvae die and turn black, a parasite wasp will emerge and continue the beneficial process. Do not throw pruned leaves away without checking them for black larvae containing parasites. Leave these under plants for about one week until wasps have emerged.

This beneficial insect is very susceptible to insecticides. It is more effective in controlling greenhouse whitefly than sweet potato whitefly. It can be purchased commercially and introduced at intervals when whiteflies are first observed. See ENT-53, Vendors of Beneficial Organisms in North America, for more information.

9.3.7 Cutworms, Armyworms, Loopers and Other Caterpillars

All caterpillars are the immature stages of moths. They chew on leaves, stems and fruits of many kinds of plants. Infestations may begin when moths enter through ventilators or when infested plants are brought into the greenhouse. Cutworms can be serious pests of younger plants. They hide during the day in soil or mulch and feed on the plants at night.

The cabbage looper can be a pest of greenhouse crops, especially lettuce. It can be distinguished by its pale green color, three pairs of prolegs (small fleshy bumps on the underside of the abdomen) and looping movement (bringing the rear of its body up to the front legs before moving the front legs forward) similar to that of measuringworms. When monitoring for these insects, look for cut plants or leaves with large sections removed. Sprays containing Bacillius thuringiensis are effective against these pests.

Figure 7. Cutworm larva

Figure 8. Cabbage looper

Leafminers

Leafminers are larvae of small flies. They damage plants by feeding between the upper and lower surface of the leaf. Damaged areas are light in color and narrow and winding. They increase in width as the larva grows. When fully grown, the larva may pupate in the leaf tissue or emerge from the leaf and fall to the gound to pupate. Each female fly will lay from 50 to 100 eggs by inserting them into pits made in the leaf surface. Because the damaging stages of the insects occur entirely inside the leaf, control with contact insecticides is ineffective once the damage appears. Infestations can be avoided through the use of good cultural practices, hand romoval and disposal of infested leaves and use of chemical controls when necessary.

Fig. 9: Leafminer larva and adult

Mealybugs

Mealy bugs are small, soft-bodied insects which, like aphids, feed on plant sap. These insects are thickly covered with mealy or waxy secretions which provide some protection from contact insecticides. Some species lay eggs; others give birth to live young. Like aphids, mealybugs often produce large amounts of honeydew that results in sooty mold on leaves and other plant parts.

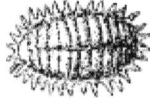

Fig. 10: Mealyong

Mealybugs may infest almost any part of the plant. A wide variety of plants in the greenhouse are susceptible to mealybugs, but they are often seen first on crotons, hoyas and bamboo palms. Ants, which collect honeydew as food, are often seen in association with mealybug infestations.

Mites

Mites are sap-sucking pests which attack a wide range of greenhouse plants. Two species, the two-spotted spider mite and the cyclamen mite, can cause serious and persistent problems. These mites feed by piercing tissue with their mouthparts and sucking out cell contents.

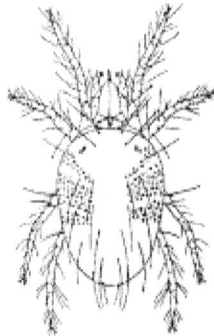

Fig. 11: Two-spotted spider mite

Two-spotted spider mites are light to dark green with two distinctive black spots on the abdomen. Eggs are spherical and clear when first laid. After hatching, the larva has three pairs of legs, but later stages will have four pairs. Males are smaller with more pointed abdomens than females. Heavy infestations of the two-spotted spider mite produce fine webbing which may cover the entire plant.

Generally, they feed on the undersides of leaves, giving the upper leaf surface a speckled or mottled appearance. Leaves of mite-infested plants may turn yellow and dry up, and plants may lose vigor and die when infestations are severe. Females can lay 200 eggs, and during hot, dry weather the life cycle may be completed in seven days. Marigolds, crotons, chrysanthemums, roses, impatiens, parlor palms, bamboo palms and ivy geraniums are highly susceptible to two-spotted spider mites and can be used as indicator plants to alert managers to infestations.

Cyclamen mites are minute, elliptical, semi-transparent, greenish mites. These mites thrive when the temperature is around 60°F and can complete their life cycle in about two weeks. African violets, cyclamen, dahlia, gloxinia and New Guinea impatiens are highly susceptible to cyclamen mites and can be used as indicator plants to alert greenhouse managers.

Depending on the type of plant attacked, cyclamen mites may infest the entire plant or be concentrated around the buds. Infested leaves become distorted and often curl inward; foliage may become darker than that of healthy leaves. Because of their small size, infestations often go undetected until the damage is severe. Usually it is the nature of the injury, not the mites themselves, that alerts greenhouse managers to cyclamen mite infestations.

Mites can easily be moved to infested plants on clothing, so always examine infested benches and other hot spots last during greenhouse inspections. Often, it is better to discard infested plants than to attempt to control the problem with pesticides. If control is attempted, isolate the infested plants to reduce potential spread.

Resistance to pesticides has increased the difficulty of controlling these pests. Because mites primarily occur on the undersides of leaves, applications of contact miticides should be directed at both the lower and upper leaf surfaces. Mite eggs are resistant to some insecticides, so repeated applications are often necessary to control infestations. Two to three applications spaced five days apart may be necessary. Miticides with different modes of action need to be alternated, so different products are used to control each subsequent mite generation.

Several species of mite predators are commercially available. These are usually released when mites first appear and should be evenly dispersed throughout the greenhouse. If mite infestations are heavy, consider spraying with an insecticidal soap before releasing predator mites. Selection of the proper predatory mite species will depend on greenhouse temperatures and humidity. If predatory mites are used, early release at the first sign of mite infestation is critical. Unlike a miticide, predatory mites will take some time to control infestations.

Slugs and Snails

Slugs and snails can become greenhouse pests when the humidity is high. Slugs are fleshy, slimy animals that feed mainly at night. They prefer cool, moist hiding places during the day. Slugs rasp on leaves, stems, flowers and roots. They produce holes in the leaves or just scar the leaf surface. Small seedlings are especially vulnerable to these creatures. Silvery slime trails are evidence of snail and slug infestations.

Sanitation is important for slug control. Keep the greenhouse free of plant debris (leaves, pulled weeds, etc.), old boards, bricks or stones that provide cool, moist hiding places for slugs. Barriers of diatomaceous earth, lime, sawdust, copper stripping and salt-embedded plastic strips can be used around benches. Metaldehyde or mesurol bait pellets can be distributed beneath the benches in greenhouses for slug and snail control on ornamentals. Do not allow pellets to come in contact with plants.

9.4 GENERAL STRATEGIES FOR INSECT AND MITE MANAGEMENT

9.4.1 Cultural Controls are Essential

Pests are generally brought into the greenhouse on new plant material. Others may enter the greenhouse in the summer when the ventilators are open. Many are able to survive short periods of time between harvest or plant removal and production of the next crop. Cultural controls are the primary defense against insect infestations.

The following cultural practices will help to prevent pest infestations:

- Inspect new plants thoroughly to prevent the accidental introduction of pests into the greenhouse.
- Keep doors, screens and ventilators in good repair.
- Use clean or sterile soils or ground media. Clean or sterilize tools, flats and other equipment.
- Maintain a clean, closely mowed area around the greenhouse to reduce invasion by pests that develop in weeds outdoors.
- Eliminate pools of standing water on floors. Algal and moss growth in these areas can be sources of fungus gnat and shore fly problems.
- Dispose of trash, boards and old plant debris in the area.
- Remove all plants and any plant debris, clean the greenhouse thoroughly after each production cycle.
- If possible, allow the greenhouse to freeze in winter to eliminate tender insects like whiteflies.
- Avoid overwatering and promote good ventilation to minimize wet areas conducive to fly breeding.
- Avoid wearing yellow clothing which is attractive to many insect pests.
- Maintain a weed-free greenhouse at all times. 12. Eliminate infestations by discarding or removing heavily infested plants.

9.4.2 Monitoring

Early detection and diagnosis of pest infestations will allow you to make pest control decisions before the problem gets out of hand. It is good practice to make weekly inspections of plants in all sections of the greenhouse. When monitoring, select plants so that they represent the different species in the greenhouse. Pay particular attention to plants near ventilators, doors and fans. At least 1% of the plants need to be examined on each monitoring visit in the greenhouse.

Insect monitoring devices should be used in the greenhouse. Yellow sticky cards (PT Insect Monitoring & Trapping System, Whitmire, St. Louis, MO) are highly attractive to winged aphids, leafminer adults, whiteflies, leafhoppers, thrips (blue cards can also be used with thrips), various flies and other insects. White sticky cards can be used to detect fungus gnat adults. These can be used to alert you to the presence of a pest and identify hot spots in the greenhouse. One to three cards per 1000 square feet in the greenhouse is recommended. Cards should be changed weekly. Typically, these sticky cards are suspended vertically just above the tops of the plants. They can be attached to sticks or hung on string. If you cannot identify a trapped insect, contact your county Extension agent for assistance.

Mass trapping products such as sticky tapes are also available for management of thrips, whiteflies, leaf miners and fungus gnats. While sticky cards are primarily used just to alert you to insect infestations, mass trapping tools are used to reduce and manage insect infestations. Mass trapping relies on using enough surface area of the attractive sticky tapes to capture and reduce pest numbers. Care should be taken to keep monitoring and trapping products dry and free of debris. This will maintain effectiveness of the traps.

9.4.3 Biological Control

Agents Natural enemies are commercially available for control of some greenhouse pests. For a listing of sources, see ENT-53, Vendors of Beneficial Organisms in North America.

Beneficial organisms commercially available for greenhouse pest management	
Beneficial organism	**Whiteflies**
Parasitic wasps, Encarsia formosa	Whiteflies
Parasitic wasps, Aphtis melinus	Scales
Leafminer parasite, Dacnusca sibiriica and Diglyphus isaea	Serpentine
	leafminers, ungus gnats
Predatory mites, Amblyseius californicus,	Spider mites
Phytoseiulus longipes and	
Phytoseiulus persimils	Thrips
Predatory mites, Amblyseius cucumeris	
and Amblyseius mckenziei	
	Various
Lady beetles, Hippodamia convergens and Cryptolaemus montrouzeri	soft-bodied insects and eggs
	Various
Green lacewings, chrysoperla carnea	soft-bodied insects and eggs

Levels of pest control obtained with beneficial organisms will vary greatly depending on a number of factors, including:

- ⊙ Species of pest involved
- ⊙ Species of natural enemy used
- ⊙ Timing of release of natural enemy relative to pest buildup and crop development
- ⊙ Numbers of beneficials released
- ⊙ Greenhouse temperature and range of fluctuation
- ⊙ Time of year
- ⊙ Condition of the beneficials at release
- ⊙ Pesticide usage before and after release of beneficials

Biological control generally requires more time than pesticides to bring a pest population under control. Natural enemies require time to disperse from release sites and to search for prey or hosts. Appropriate natural enemies should be released as soon as the pest is detected in the greenhouse. Natural enemies do not provide sufficiently rapid control of pests that are already causing serious losses, and they will not generally eradicate an infestation. In some instances, using an insecticidal soap or other non-residual insecticide is recommended to reduce the infestation before releasing the natural enemies. Knowledge of pest biology and monitoring of pest populations are critical to determining when to make releases. Greenhouse managers should avoid unnecessary insecticide/miticide applications before and after release of natural enemies. If insecticide/miticide treatments are required, limit treatments to pest "hot spots" to avoid treating the entire greenhouse. Use a selective, short residual pesticide if possible. For example, *Bacillius thuringiensis* (Bt) products can be used to control caterpillars without harm to natural enemies in the greenhouse.

9.4.4 Pesticide Management

Greenhouse operators need to maximize the effectiveness of insecticides and miticides. To provide adequate control, a pesticide must be applied at the proper rate, when the pest is present. Coverage and sufficient pressure are needed to penetrate dense foliage and reach the target pest. This is especially important for sucking insects that infest the lower surface of leaves. Older, lower leaves can be removed to open the canopy of some crops to increase spray coverage. Insecticide or miticide applications must sometimes be repeated frequently to maintain a pest at acceptable levels. Timing of pesticide applications is important. Some pests are vulnerable to pesticides only at certain stages in their life cycle. For whitefly management, begin control measures early. If control action is delayed until an abundance of adult whiteflies can be seen, then numerous eggs and immature stages, which are more difficult to control, are usually present.

With a limited number of pesticides available for greenhouse use, it is always a concern that pests may develop resistance to pesticides. Managers should rotate among different pesticides for successive applications when controlling specific pests. Rotations must include pesticides belonging to different chemical classes that use different modes of action to control the pests. This will prevent, or at least delay, the development of resistance to a particular pesticide.

To aid pesticide applications, plants that are frequently infested by the same pest and can be legally sprayed with the same material should be grouped together. This will reduce the potential for misapplications to unlabelled crops. Additionally, moving infested material through the greenhouse can spread an infestation to other areas.

10

INSECT PESTS OF STORED GRAIN AND THEIR MANAGEMENT

10.1 INTRODUCTION

The losses caused by the storage pests which include, insects, fungi, weeds, rodents and abiotic factors is to the tune of 10% (FAO, 1978). It has been estimated that between one quarter and one third of the world grain crop is lost each year during storage. Much of this is due to insect attack. In addition, grain which is not lost is severely reduced in quality by insect damage. Many grain pests preferentially eat out grain embryos, thereby reducing the protein content of feed grain and lowering the percentage of seeds which germinate. Some important stored grain pests include the lesser grain borer, rice weevil and rust red flour beetle.

Overseas customers demand insect-free grain. For this reason, the Australian Department of Agriculture has imposed nil tolerance of insects in export grain. Insect pests also increase costs to grain growers both directly through the expense of control on the farm, and indirectly through the costs incurred by grain handling authorities in controlling weevils in bulk storages.

Grain insect pests may be divided into primary and secondary pests. Primary grain insects have the ability to attack whole, unbroken grains, while secondary pests attack only damaged grain, dust and milled products.

10.2 CATEGORY OF STORAGE PESTS

Storage insect pests are categorized into two types:

10.2.1 Primary storage pests:

Primary storage pests are those insects that damage sound grains or Primary storage grain insects pests have the ability to attack whole, unbroken grains. The eggs are laid outside the grain, before the larvae mature inside the grain and then chew their way out.

Primary storage pests are categorized into two type, Internal and External feeders/borers.

INTERNAL FEEDERS/BORERS:

S. No.	Name of Insect	Scientific Name of Insect	Family and Order of Insect	Damaging stage of Insect
1.	Rice Weevil	Sitophilus oryzae	Curculionidae: Coleoptera	Grubs and adults
2.	Pulse beetles	Callosobruchus chinensis, C. maculates, C. analis	Bruchidae : Coleoptera	Grubs
3.	Lesser/Hooded/ Paddy grain borer	Rhyzopertha dominica	Bostrychidae : Coleoptera	Grubs and adults
4.	Potato tuber moth	Phthorimaea operculella	Gelechiidae : Lepidoptera	Caterpillar
5.	Angoumois grain moth or grain moth	Sitotroga cerealella	Gelechiidae : Lepidoptera	Caterpillar
6.	Sweet potato weevil	Cylas formicarius	Apionidae : Coleoptera	Grubs and adults
7.	Tamarind beetle	Pachymeres gonagra	Bruchidae : Coleoptera	Grubs and adults
8.	Drug store beetle	Stegobium panaceum	Anobiidae : Coleoptera	Grubs and adults
9.	Cigarette beetle	Lasioderma serricorne	Anobiidae : Coleoptera	Grubs and adults
10.	Groundnut bruchid/ Tamarind bruchid or Peanut bruchid	Caryedon serratus	Chrysomelidae : Coleoptera	Grubs
11.	Arecanut beetle	Araecerus fasciculatus	Anthribidae : Coleoptera	

External Feeders/Borers:

S. No.	Name of Insect	Scientific Name of Insect	Family and Order of Insect	Damaging stage of Insect
1.	Rust Red flour beetle	Tribolium castaneum, T. confusum	Tenebrionidae: Coleoptera	Grubs and adults
2.	Khapra beetle/ Warehouse beetle	Trogoderma granarium	Dermestidae : Coleoptera	Grubs
3.	Indian meal moth	Plodia interpunctella	Phycitidae : Coleoptera	Caterpillar
4.	Fig moth or Almond moth or Warehouse moth or Dried currant Moth	Ephestia cautella This is serious pest of Walnut/Raisin/Dates/ Berries	Phycitidae : Lepidoptera	Caterpillar
5.	Rice moth	Corcyra cephalonica	Galleridae : Lepidoptera	Caterpillar

Primary Grain Pests

Lesser grain borer (*Rhyzopertha dominica*): The lesser grain borer is the most serious pest of stored grain in Western Australia. It is a dark brown cylindrical beetle about 3mm long. The head is hidden by the thorax when viewed from above. Females lay up to 500 eggs scattered loosely through the grain. The eggs hatch to produce curved white larvae with brown heads and three pairs of legs. The larvae burrow into slightly damaged grains and eat out the starchy interior. After pupating the adults emerge from the grain, leaving large irregular exit holes. The life cycle takes from 3-6 weeks depending on the temperature. Adults may live up to two months.

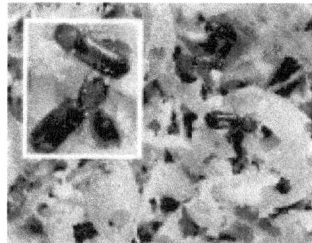

The adult lesser grain borers chews grain voraciously causing damage which may facilitate infestation by a secondary pest. It is a strong flyer and may rapidly migrate from infested grain to begin new infestations elsewhere.

Granary weevil (*Sitophilus granarius*)

with six tooth-like projections on each side of the thorax. They lay up to 500 eggs loosely spread through the infested grain; eggs hatch to produce larvae which feed externally on grain dust and sometimes wheat embryos. The mature larvae pupate within a silken cocoon. A complete generation may take place in as little as three weeks but the adults may live up to nine months. They frequently hide in cracks and crevices of buildings and machinery.

Flat grain beetle (Cryptolestes spp.)

Flat grain beetles are small reddish-brown insects about 1.5mm long with long antennae and a flattened body. Eggs are laid throughout the stored grain and develop into tiny larvae with characteristic tail horns, biting mouth parts and three pairs of legs. They feed on damaged grain and wheat embryos. Pupation takes place in a cocoon. A complete life cycle takes from 4-5 weeks and adults may survive up to one year.

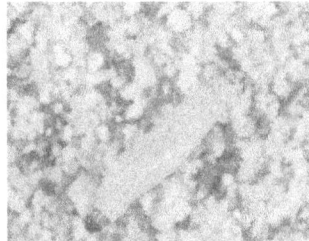

Indian meal moth *(Plodia interpunctella)*

The adult Indian meal moth is grey with distinctive brownish-red tips to the forewings. The female lays up to 200 eggs near the grain surface as it slowly passes from grain to grain spinning a silk thread. Severe infestations may form a surface web on the grain heap. Larvae attack the wheat germ, then pupate in a cocoon which may be found in cracks and crevices of buildings. The insects quickly emerge as adult moths. A generation takes as little as four weeks under warm conditions.

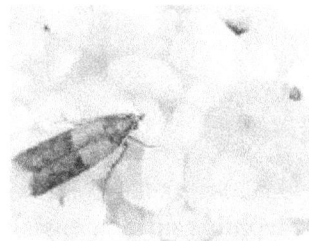

Warehouse beetle (*Trogoderma variable*)

This beetle was first found in WA in 1989 and that infestation was eradicated. It has now spread to about a dozen locations in WA but has never established to damaging populations.

The warehouse beetle is a pest of stored grain in its own right, but the greater threat is the impact on trade that it could have by masking an incursion of the world's worst pest of stored grain — the khapra beetle. Warehouse beetle and khapra beetle require microscopic examination to distinguish them. Khapra beetle does not occur anywhere in Australia and would have a severe impact on international trade if it became established.

Eggs are usually laid in crevices and under the surface of loose food. They hatch in about a week. Only the larval stage damages grain. It is frequently found in seeds, groceries and used sacks. The larvae are conspicuously hairy. They usually live for about five weeks but may enter a dormant phase (diapause) for more than two years. Larvae may moult up to ten times.

After pupation adults emerge. They are less obvious than the larvae and do no damage to grain. They live for up to five weeks during which females lay up to 80 eggs. Warehouse beetles cannot fly and are spread only in infested commodities and old sacks.

A characteristic of warehouse beetle infestations is the accumulation of cast larval skins. Hairs shed by larvae may cause asthma, skin or gastric problems.

It is impossible to distinguish between T. variabile and several harmless native species without the aid of a microscope. Any hairy larvae found in grain stores should be sent to DPIRD for positive identification.

10.3 IMPORTANT POINTS TO BE REMEMBERED:

⊙ Khapra beetle is native to India.

⊙ Pulse beetle prefers whole pulse not split pulse.

⊙ Use fumigants like ethylene dibromide (EDB), ethylene dichloride carbon tetra chloride (EDCT), aluminium phosphide (ALP) to control stored produce pests effectively.

⊙ Granary weevil (Sitophilus granarius, the Wheat weevil).

⊙ Crops can be completely destroyed or even partially damaged affecting the quality of the crop and the ability to germinate new ones, by decreasing the protein content and removing the seeds from the grains.

- Flat grain beetle is shortest insect in stored grain pests.

- Gaseous quinones released to the medium produces a readily identifiable acid odour in heavy infestations of - Red flour beetle: Tribolium castaneum

- Caryedon serratus is longest insect in stored grain pests.

- Name a storage pest on arecanut, coffee and cocoa - Arecanut beetle: Araecerus fasciculatus

- Caryedon serratus is a major stored grain pests of groundnut.

- Long headed flour beetle: Latheticus oryzae resembles – Tribolium castaneum

- Caterpillars of Angoumois grain moth produce large quantities of silk webbing.

- Sweet potato weevil and Potato tuber moth infestation is observed both in field and storage.

- Sitophilus oryzae, Sitotroga cerealella, Caryedon serratus, Tamarind beetle and Callosobruchus sp. Infestation starts from the field itself and carried to the godowns.

- A storage grain mixed with Malathion WP.

- For the long term safe storage of seeds the moisture content of the seed should be 7-8%.

- Chlorpyrifos methyl (Reldan) is effective against stored grain pests Khapra beetle, Rust Red flour beetle and Sitophilus oryzae but except stored grain pest Rhyzopertha dominica.

- Stored grain pest which is more sensitive /susceptible to spinosad is Rhyzopertha dominica.

- Potato tuber moth controlled by egg/larval parasitoid Chelonomus blackburni.

- The first case of insecticide resistance in India in stored grain pest was Tribolium castaneum.

- Chlorpyrifos methyl (Reldan), Cyfluthrin (Tempo) and Pirimiphos methyl (Actellic) insecticides used for the management of stored grain pests.

- Pusa bin is used for the control of Trogoderma.

- One Aluminium Phosphide tablet (three grams) release one gram of Phosphine gas.

- Hidden infestation can be detected by Staining method, Ninhydrin method and KOH method.

- The dropping are spindle shaped in Roof rat.

- Value depreciation of a product owing to the present of insects or other foreign matter is known as Dockage.

- Infestation of most of the insect pests can be avoided if the moisture content is below 9 % except Khapra beetle insect.

- Impact machine which are used to kill insects in stored products are called as Entoletors.

- The composition of Aluminium Phosphide is Aluminium Phosphide (56%), Aluminium Carbamate (41%) and Paraffin wax (3%).

- The concentration of the gas per unit volume x time of exposure is called as Critical value (CT Value).

- Phosphine gas that is liberated when Aluminium Phosphide tablet reacts with moisture.

- Recommended dose of Phosphine fumigation is 3 gram Aluminium Phosphide tablet/tonne with an exposure period of 7 days.

- Zinc phosphide and Barium carbonate is a acute rodenticides (Single dose and quick acting)

- Warfarin and Bromodiolone is a Chronic rodenticides (Multiple dose and slow acting)

- Bromodiolone is Multiple dose poision, slow acting poision and anticoagulant.

- Methyle Bromide, Sulfuryl floride (Profume) and Phosphine fumigant used for the management of stored grain pests.

- Grain Psocids (Liposcelis spp) are scavengers and most common in slightly damp stores/materials.

- Storage pest that feeds on animal products (dried blood & dried insect) is known as Khapra Beetle.

- Khapra beetle hibernates in cracks and crevices during grub stage.

- Khapra beetle adult does not consume the grains but only grub is damaging stage.

- Saw toothed grain beetle grub pupate within a silken cocoon.

- Saw toothed grain beetle hide in cracks and crevices of building and machinery.

- The Saw toothed grain beetle name has been derived due to characteristics teeth running down the pro thorax(Six tooth like projections on each side of thorax)

- Insect pests attacking seeds in storage lower the percentage of seeds which germinate.

- Many grain pests preferentially eat out grain embryos, thereby reducing the protein content of feed grain.

- Pulse beetle hibernation in grub stage.

- Adult of pulse beetle does not feed seed.

- Angoumois grain moth besides reducing the grain weight, impart an unpleasant smell and taste to the cereals.

- The storage pest that is commonly found in broken cereals and their milled products like atta, maida and suji is known as Rust Red Flour Beetle

- Flate grain beetle pupation take place in a cocoon.

- Lesser grain borer is most serious in hot and dry conditions.

- Atta formation is the main symptoms of Lesser grain borer damage by which infestation in the godown can be easily detected.

- Unlike the Lesser grain borer (A strong flyer and may rapidly migrate from infested grain to begin new infestations elsewhere), the rice weevil may occasionally fly.

- Rice weevil pupate inside the grain.

- Indian meal moth female lay eggs near the grain surface.

- Red Squill and Strychnine is a rodenticide of botanical origin.

- Khatti is an underground storage structure.

Insecticides approved by Registration Committee (RC) for control of stored grain pests under the Insecticides Act, 1968.

- Aluminium Phosphide

- Deltamthrin 2.5% WP

- Methyle Bromide

- Ethylene Dichloride + Carbon Tetrachloride (3:1)

- Methyle Bromide (98%) + Chloropicrin 2 % w/w.

Phosphine fumigation

- Larval and adult stage are more susceptible than egg and pupa.

- It has little reaction with more commodities,

- It does not effect the germination and milling qualities.

Methyle Bromide fumigation

- All insect stages are effective and requires less exposure period.

- It is widely used for plant quarantine and food processing facilities.

- It destroys stratospheric ozone layer.

Anticoagulant

- They are safer than acute rodenticides because they are less toxic to the non target species.

- ⦿ Quick knock down effect.
- ⦿ Cost of operation is cheap.

Disadvantage of Zinc phosphide baiting is:

- ⦿ Necessity of prebaiting
- ⦿ Induce bait shyness
- ⦿ Toxix to non target species.

Table: Insects Pests with their Characteristic/Damaging/ Typical/ Identified Symptom

Sl. No.		Characteristic/Damaging/ Typical/ Identified Symptom
1.	Almond Moth	Presence of silken tubes
2.	Rice Moth	Webbing of Rice Grains
3.	Lesser grain borer	Irregular holes on the grains
4.	Rice Weevil	Circular bore hole in rice
5.	Pulse beetle	Circular hole or exit hole
6.	Angoumoise grain moth	Grain covered with scales
7	Angoumoise grain moth	Caterpillars produce large quantities of silk webbing
8.	Indian meal moth	Damaged grains and webbing
9.	Red flour beetle	Powdery grains and foul smell
10.	Tamarind beetle	Circular holes on fruits
11.	Cigarette beetle	Processed tobacco is presence of round pin head sized bore holes

10.4 MANAGEMENT OF STORED INSECTS

It is broadly classified into two categories, preventive and curative methods.

10.4.1 Preventive or Prophylactic method:

- ⦿ Clean godown before storing the harvested crop.
- ⦿ Seal and plaster the cracks and crevices in godowns that are hiding places for most of the storage insect pests including rodents.
- ⦿ Machineries should be properly cleaned before being used for processing and threshing.
- ⦿ To avoid entry of birds, rats and squirrels, fix wire meshes to windows, ventilation etc.
- ⦿ Drying of seeds under natural sunlight exposes that larva or eggs on grains and kill them.
- ⦿ Storing the seeds under good ventilation godowns and warehouses to prevent build up of moisture which is favorable to insects for multiplication. For example rice can be safely stored at 13 % moisture content whereas wheat, sorghum and

maize may be stored safely at 13.5 % moisture content at a temperature of 27°C as given in below table

Paddy	15*
Rice	13
Wheat, sorghum and maize	13.5
Groundnut (shelled)	7
Mustard seeds	5-6 %
Cowpea/ beans	15

* Value are upper limit values.

Spraying of insecticides on godown walls , floor, alleyways and surface of the grain bags with **Malathion 50 % ECV(1:100 dilution rate) and Deltamethrin 2.5 % WP (40 grams in 1 liter of water)** @ 3 liters emulsion on 100 m² surface area.

10.4.2 Curative/Physical Methods

⊙ Additional methods of insect control include smoking, sun-drying, admixing of dusts or ashes or mixing of dusts or ashes with seeds.

⊙ Use of insecticides like Malathion, Dichlorvas, and Deltamethrin as prophylactic spray on godown walls, storage surface etc.

⊙ Use of fumigants viz., Aluminium phosphide sold under trade name Celphose(3tablets/tonne of grain) that is properly used to fumigate the grains. One Aluminium phosphide tablet is of 3 gram weight which release one gram of phosphine gas.

⊙ Methyle bromide (98%) is also used for fumigation against whole cereals, millets and pulses at 24g/m³ in air tight cover with 6-8 hrs exposure period and waiting period of 24 hrs.The residues should not exceed 25 ppm and similarity it is approved for use of milled products (flour) and also for dry fruits, nuts, spices and oil seeds @ 24-32 g/m³ with 12-24 hrs exposure and waiting period of 72 hrs.

⊙ Mixing of seeds only with Malathion WP.

⊙ Once grains are infested , even controlled atmosphere technique can be used to kill storage pests where the gases viz., CO_2 concentration is increased in the close spaces which close the spiracles of insects and kills the insects through asphyxiation(Physical control).

⊙ The minimum exposure period of fumigation in godowns is 5-7 days except for empty godowns and sheds it is 3 days.

• **Fumigation dose based on Stored quantity:** Fumigation with Aluminium phosphide 56% for stored whole cereals, seed grains, millets, pulses, dry fruits, nut, spices & oilseeds, milled products, deoiled cakes rice bran flour, grain animal & poultry feed, split pulses (Dal) and other processed foods etc @ 3 tablet/ ton.

- **Fumigation dose based on Space occupied:** Fumigation with Aluminium phosphide 56% for stored whole cereals, seed grains, millets, pulses, dry fruits, nut, spices & oilseeds is **150g/100m³**

- **Whereas** milled products, deoiled cakes rice bran flour, grain animal & poultry feed, split pulses (Dal) and other processed foods is **225g/100m³**

- **Fumigation dose for Empty Godowns & Sheds:** Fumigation with Aluminium phosphide 56% for Empty Godowns & Sheds is **14 tablets/1000 Cu ft. or 150g/100m³**

10.5 INSECTICIDE RESISTANCE

- ◉ Insect populations of many species have evolved resistance to insecticides as a result of the widespread use of these chemicals in control. In some cases, insects which have only been exposed to one insecticide develop resistance to other, related compounds.

- ◉ It takes many years and millions of dollars to develop and test new compounds. Therefore, it is important that insecticide resistance is prevented from spreading. This may be achieved by appropriate use of pesticides and by farm hygiene. This consists of careful cleaning of all machinery and buildings used for storing and transporting grain right from the header to the port terminal.

RELATED TERMINOLOGY

Abdomen: The third or posterior division of the insect body

Abdominal: Part of the abdomen

Antennae: Paired sensory organs originating on the insect head

Antennal Article: An individual segment of an antenna

Antennal insertion: The point where an antenna attaches to the head

Anterior: In front; the front of

Anterior angle: The angle of the thorax near the head

Apex: The portion of a body part farthest from the base or point of attachment. The apex of the elytra is the portion at the rear end of the elytra.

Binomial: Consisting of two parts

Canthus: Ridge dividing the eye of some insects into an upper and a lower half

Carina: An elevated ridge

Club: Portion of the antenna that is enlarged from the other segments

Coleoptera: The order comprised of the beetles; sheath winged

Concave: Hollowed out; a depression

Confluent: Running together

Convex: Curved outward; opposite of concave;

Crenulate: With small scallops, evenly rounded

Declivity: Sloping downward

Diapause: A condition of suspended animation; no activity or development occurs

Dorsal: Upper surface or back when viewed from above

Dorsum: In general, the upper surface

Elytra: First pair of wings that are modified to form a hard shell

Facet: One portion or segment of the compound eye

Femur: The thigh; usually the stoutest segment of the leg

Filiform: Thread like

Frons: The upper, anterior portion of the head

Fungal feeder: Insect that feeds exclusively on fungus associated with stored grain

Genera: The name of a genus, the first portion of a binomial scientific name

Grub: An insect larva in the Order Coleoptera

Head: The first or anterior division of the insect body where the eyes and antennae are found

Horns: Pointed process of the head

Hypopus: Resting larval stage of certain mite species

Incision: Any cut into a margin or through a surface

Incision: Outer covering or cuticle of the insect body

Interval(s): Space between two structures or sculptures e.g. space between two ridges on elytra

Larval: Pertaining to immature stage of juvenile insects

Lateral: Pertaining to the side

Lateral: Bead edge of pronotum slightly to moderately thickened forming a bead

Lateral: Ridges a raised line along the side of a portion of the insect

Mandibles: First pair of jaws in insects

Mandibles: In or at the middle

Mesothorax: Second or middle segment of the thorax bearing the second or middle pair of legs and the anterior or first pair of wings

Metathorax: The third or last segment of the thorax bearing the third or hind pair of legs and the second pair of wings

Morphological: Relating to form and structure

Moult: Process of larval growth involving the shedding of outgrown skin

Ocellus: Simple eye in adult insects consisting of a single bead like lens

Pits: Same as punctures; small impression on the hard outer part of the insect body

Posterior: Hindmost; at the back

Primary Insect: Ss defined by the Canadian Grain Commission, an insect that can attack whole, sound grain and requires action

Pronotum: The upper or dorsal surface of the prothorax; on the beetles this appears to be the middle segment when viewed from above

Prothorax: The first segment of the thorax bearing the first pair of legs

Pubescence: Short, fine, soft hair

Punctures: Small impression on the hard outer part of the insect body

Puparia: Thickened, hardened barrel-like larval skin within which the pupa is formed

Scavenger: A feeder on decaying or waste matter

Secondary Insect: An insect that requires grain that is going out of condition or damaged to be able to feed on it

Serrate(d): Saw-like or toothed

Seta: Single hair

Setae: Hairs

Snout: Lengthening of the head to give the appearance of a nose or snout

Striated: Marked with fine, parallel, impressed lines

Sub-Basal: Below the base or point of attachment

Subequal: Similar, but not equal in size

Sublateral: Nearly to the side

Subparallel: Nearly parallel

Temple: Region between the eye and the back of the head

Terga: Belonging to the upper surface of the body (more than one segment)

Tergum: The upper or dorsal surface of any one body segment

Thoracic Shield: Upper surface of the thorax that appears as a single segment and is shield-shaped

Tubercle: Small bumps or projections from a surface

Urogomphi: Fixed or mobile structures found on the last or terminal segment of certain larvae

Ventral: The under surface of the abdomen; from below

Aedeagus: A tube-like organ which enters the female's body during copulation (like a penis)

Aeropyles: Microscopic pores in the chorion that allow respiratory exchange of oxygen and carbon dioxide with relatively little loss of water.

Ametabolous: Adjective - refers to insects where adults and nymphs are wingless, and there is no visible change in form between the stages, other than in size. Ametabolous development occurs in the wingless insects apterogytes and in other groups that undergo simple metamorphosis where the adults are wingless (1). ametabola refers to this group of insects (sometimes called apterogytes), includes: Zygentoma - Silverfish Microcoryphia - Bristletails

Apodeme: A ridge-like ingrowth of the exoskeleton that supports the internal organs and provides the attachment points for the muscles.

Apolysis: Physical separation of the epidermis from the old endocuticle during molting.

Apophyses: Finger-like invaginations of exoskeleton; an elongate apodeme (an internal projection of the exoskeleton)

Blood Sinus: A body cavity where the blood flows in the insect's open circulatory system.

Bursa Copulatrix: The part of the female genitalia which receives the aedeagus and sperm during copulation

Campodeiform: Larval body type often called crawlers. Elongated, flattened body with prominent antennae and/or cerci; thoracic legs adapted for running.Example: lady beetle larva

Cement layer: Layer of epicuticle that protects the wax layer from heat or abrasion.

Cerci: Paired sensory appendages on the rear-most segments of some "primitive" insects. In some cases, they may also serve as weapons or copulation aids.

Chorion: The protective "shell" of protein covering the insect's egg that is secreted before oviposition by accessory glands in the female's reproductive system.

Chrysalis: Pupal stage of holometabolous insects in which developing appendages (antennae, wings, legs, etc.) are held tightly against the body by a shell-like casing. Often found enclosed within a silken cocoon. Ex. Butterflies and moths

Cibarium: The portion of the preoral cavity between the hypopharynx and the labrum cleavage energids In an embryo, the daughter nuclei cleavage products and their surrounding cytoplasm.

Clypeus: Facial sclerite located just below the epistomal suture

Coarctate: Pupal body type. Body is encased within the hard exoskeleton of the next-to-last larval instar. Example: Flies

Collophore: A fleshy, peg-like structure found in Collembola on the ventral side of the first abdominal segment. It appears to maintain homeostasis by regulating absorption of water from the environment.

Corona: The dorsal, anterior region of the head capsule; equivalent to the forehead.

Coronal Suture: Suture along dorsal midline of head; runs backwards from the vertex along the top of the head

Ccoxa: The basal segment of the leg, by means of which it is articulated to the body

Cuticulin Layer: The innermost layer of epicuticle that is composed of lipoproteins and chains of fatty acids embedded in a protein-polyphenol complex.

Ecdysis: Shedding the old exo- and epicuticle

Eclosion: The emergence of an adult insect from its pupal case, or the hatching of an insect larva from an egg. From the French eclosion, from eclore, to open.

Ectoderm: In embryonic development, the germ layer giving rise to the epidermis, exocrine glands, brain and nervous system, sense organs, foregut and hindgut, respiratory system, and external genitalia.

Elateriform: Larval body type often called wireworms. Long, smooth, and cylindrical body with hard exoskeleton and very short thoracic legs. Example: Click beetle, flour beetle

Endocrine Glands: Secretory structures adapted for producing hormones and releasing them into the circulatory system

Endocrine System: A group of hormone-secreting structures that help maintain homeostasis, coordinate behavior, and regulate growth, development, and other physiological activities

Endoderm: In embryonic development, the germ layer giving rise to the midgut.

Epicuticle: The epicuticle is the outermost part of the cuticle. Its function is to reduce water loss and block the invasion of foreign matter.

Epiproct: The epiproct is the last dorsal sclerite at the tip of the abdomen. It covers and protects the anus from above.

Epistomal Suture: The epistomal suture runs along the ventral side of the frons. It separates the frons from the clypeus

Eruciform: Larval body type often called caterpillars. Cylindrical body with short thoracic legs and 2-10 pairs of fleshy abdominal prolegs. Example: Moths and butterflies

Exarate: Pupal body type. All developing appendages are free and visible externally. Example: Beetles and lacewings

Femur: The long leg segment between the trochanter and tibia

Flagellomere: The three basic segments of the antenna are the scape (base), the pedicel (stem), and the flagellum, which is comprised of units known as flagellomeres.

Flagellum: Long, apical segment of the antenna, commonly subdivided into several subsegments (flagellomeres)

Frons: The frons is the front of the face. It lies between the frontal sutures and above the epistomal suture.

Frontal Sutures: Frontal sutures separate the frons from the gena.

Furca: A special "strut" of exoskeleton that reinforces the ventral corners of each thoracic segment and provides a rigid site for attachment of leg muscles and ventral longitudinal muscles.

Furcula: The furcula ("little fork" in Latin) or wishbone is a forked bone found in birds and some other animals, and is formed by the fusion of the two clavicles. In birds, its primary function is in the strengthening of the thoracic skeleton to withstand the rigors of flight.

Gastrulation: A phase early in the development of the embryo during which the morphology of the embryo is reorganized to form the three embryonic germ layers: ectoderm, mesoderm, and endoderm. Each layer gives rise to specific tissues and organs in the developing embryo.

Gena: The lateral "cheek" region of the head that lies behind the frontal sutures on each side of the head

Germ Band: Blastoderm cells on one side of the egg begin to enlarge and multiply. This region, known as the germ band (or ventral plate), is where the embryo's body will develop.

Hemimetabolous: Hemimetabolous insects exhibit gradual changes in body form during morphogenesis. Immatures are called nymphs or, if aquatic, naiads.

Holometabolous: insects which pass through a complete metamorphosis in which the larva is very different from the adult and does not become more like the adult, but transforms dramatically by means of a pupal stage

Imaginal Discs: Latent adult structures in an immature insect which are clusters of undifferentiated (embryonic) tissue that form during embryogenesis but remain dormant throughout the larval instars.

Imago: An insect is known as an imago (adult) when it becomes sexually mature. At this point, molting stops and energy for growth is channeled into production of eggs or sperm.

Instar: The growth stage between two successive molts.

Labium: The labium is the most posterior of the insect's mouthparts. Its sclerites are fused along the midline to form a back lip. A pair of labial palps are sensory in function.

Labrum: The labrum is the most anterior of the insect's mouthparts. It is a flat sclerite that serves as a front lip.

Malpighian Tubules: A multitude of long, spaghetti-like structures that extend throughout most of the abdominal cavity where they serve as excretory organs, removing nitrogenous wastes (principally ammonium ions, NH4+) from the hemolymph. The toxic NH4+ is quickly converted to urea and then to uric acid by a series of chemical reactions within the Malpighian tubules.

Mandibles: A pair of highly sclerotized, unsegmented "jaws" located between the labrum and maxillae

Maxillae: Pair of mouthparts located just behind the mandibles. They help manipulate the food and include a pair of maxillary palps that are sensory in function.

Mesoderm: In embryonic development, the germ layer giving rise to the heart, blood, circulatory system, muscles, endocrine glands, fat body, and gonads (ovaries and testes).

Micropyle: A special opening near the anterior end of the chorion that serves as a gateway for entry of sperm during fertilization.

Molting: The periodic formation of new exoskeleton, often accompanied by structural changes in the body wall and other organs, followed by ecdysis (the shedding of old exoskeleton)

Naiad: The immature instar of an aquatic hemimetabolous insect

Neurohemal Organs: Specialized organs, similar to glands, that store their secretory product in a special chamber until stimulated to release it by a signal from the nervous system (or another hormone).

Neurosecretory Cells: Specialized nerve cells (neurons) that respond to stimulation by producing and secreting specific chemical messengers. Functionally, they serve as a link between the nervous system and the endocrine system.

Notum: Dorsal region of a thoracic segment

Obtect: Pupal body type called a chrysalis. Developing appendages (antennae, wings, legs, etc.) are held tightly against the body by a shell-like casing. Often found enclosed within a silken cocoon. Example: Butterflies and moths

Occiput: A sclerite that circles the foramen magnum and forms the posterior surfaces of the head capsule

Ocelli: Three "simple" eyes, one adjacent to each compound eye and one in the central groove

Ootheca: Production of an öotheca is a special adaptation found only in the cockroaches and praying mantids. The female's reproductive system secretes a special capsule around her eggs. This structure, known as an öotheca, may be dropped on the ground, glued to a substrate, or retained within the female's body.

Ovipary: The type of reproduction occurring in most insects in which life begins as an independent egg.

Ovipositor: The female's ovipositor is used to lay eggs. It consists of two large pairs of valvulae for digging into the soil and a smaller pair of valvulae for manipulating the egg during oviposition.

Paramere: Genital appendages of the male that are divided into the external and internal mera (parts).(plural paramera)

Paraproct: Paired sclerites on each side of the epiproct. They cover and protect the sides of the anus.

Pedicel: The second antennal segment, between the scape and the flagellum

Periplasm: The egg cell's cytoplasm is usually distributed in a thin band just inside the vitelline membrane (where it is commonly called periplasm).

Peritrophic Membrane: A semipermeable membrane lining the midgut that protects the delicate digestive cells without inhibiting absorption of nutrient molecules. It consists of chitin fibrils embedded in a protein-carbohydrate matrix.

Phallobase: In male genitalia, the support for the aedeagus (the male copulatory organ)

Pile: Tiny hair-like projections or surface sculpturing of the cuticle are known as microtrichae or pile. These acellular structures consist of a solid core of exocuticle covered by a thin layer of epicuticle.

Pleural Wing Process: In thoracic segments that bear wings, the pleural apodeme runs dorsally into the pleural wing process, a finger-like sclerite that serves as a pivot or fulcrum for the base of the wing.

Pleuron: The lateral plate of each insect segment -- it is usually divided by a pleural suture into at least two sclerites:; an anterior episternum and a posterior epimeron.

Pretarsus: The terminal segment of the tarsus and any other structures attached to it.

Procuticle: A multi-layered region of the exoskeleton that lies immediately above the epidermis. It contains microfibers of chitin surrounded by a matrix of protein that varies in composition from insect to insect and even from place to place within the body of a single insect.

Prothoracic Glands: A pair of endocrine glands found in the prothorax (behind the head) that manufacture ecdysteroids ("molting hormones")

Raptorial: Legs adapted for catching and holding prey

Rectal Pads: Six organs embedded in the walls of the rectum that efficiently facilitate the recovery of more than 90% of the water from a fecal pellet before it passes out of the body through the anus.

Saltatorial: Legs adapted for jumping

Scale: A flattened seta (hair); often pigmented. Characteristically found covering the body and wings of Lepidoptera.

Scape: The basal segment of an insect antenna that articulates with the head capsule

Scarabaeiform: Larval body type often called white grubs. Robust and "C"-shaped body with no abdominal prolegs and short thoracic legs. Example: June beetle and dung beetle

Sclerites: Rigid "plates" of exoskeleton surrounded by membrane or sutures are known as sclerites.

Sclerotization: Hardening of the cuticle; sclerites harden and darken as quinone cross-linkages form within the exocuticle. This process (also called tanning) gives the exoskeleton its final texture and appearance.

Serosa: Cells in the blastoderm become part of a membrane (the serosa) that forms the yolk sac. Cells from the serosa grow around the germ band, enclosing the embryo in an amniotic membrane.

Setae: Larger hairs, bristles, and scales (called setae or macrotrichae) that project from the integument. (sing. seta)

Spiracles: Spiracles are valve-like openings in the exoskeleton that regulate the flow of air into and out of the tracheal system. Spiracles are located laterally along the thorax and abdomen of most insects -- usually one pair per body segment.

Spurs: Multicellular projections of the exoskeleton are called spines (or spurs, if movable). They are lined with epidermis and contain both procuticle and epicuticle.

Sternite: Simple sclerite that covers the ventral surface of each abdominal surface; a subdivision of a sternum.

Sternum: Ventral thoracic sclerites are called sterna (or sternites).

Stigma: Rectangular stigma (pigmented patch) near tip of each wing of dragonflies and damselflies

Stylet: One of the elongate parts of piercing-sucking mouthparts; a needle-like structure.

Subgenital Plate: The subgenital plate closes over and protects the male genitalia. The male's aedeagus is usually hidden away under the subgenital plate.

Subimago: Winged "subadult" that undergoes one molt and becomes the adult; unique to mayflies (Ephemeroptera)

Suture: An external shallow groove at the junction between two sclerites.

Tagma: In biology a tagma (Greek: τάγμα, plural tagmata - τάγματα) is a specialized grouping of multiple segments or metameres into a coherently functional morphological unit. Familiar examples are the head, the thorax, and the abdomen of insects.

Tarsomere: A subdivision of the tarsus (the leg segment distal to the tibia)

Tarsus: Distal, segmented part of the insect leg attached to the tibia; usually subdivided into 1-5 tarsomeres

Tenaculum: A minute structure on the sternum of the third abdominal segment which serves as a clasp for the furcula of collembolans

Tentorium: Inside the head, the tentorium serves as an internal "truss" that reinforces the head capsule, cradles the brain, and provides a rigid origin for muscles of the mandibles and other mouthparts.

Tergum: Simple sclerite that covers the dorsal surface of each abdominal segment

Thorax: Middle portion of the body between the head and abdomen, consisting of three segments (prothorax, mesothorax, and metathorax), each of which usually bear a pair of articulated legs

Tibia: Long leg segment located between the femur and the tarsus

Tormogen Cell: The socket-forming epidermal cell associated with a seta.

Tracheal System: A complex network of tubes that delivers oxygen-containing air to every cell of the insect's body.

Trichogen Cell: A hair-forming epidermal cell associated with a seta.

Trochanter: Small segment of the leg that articulates with the coxa and forms an immovable attachment with the femur

Valvifers: Basal sclerites with muscle attachments that support the valves of the ovipositor.

Valvulae: Apical sclerites of the ovipositor which guide the egg as it emerges from the female's body.

Vermiform: Larval body type often called maggots. Fleshy, worm-like body with no head capsule or walking legs. Example: House fly and flesh fly

Vertex: The vertex is the forehead - - where the head narrows to a point.

Vitelline Membrane: An egg's cell membrane made up of a phospholipid bilayer similar in structure to most other animal membranes. It surrounds the entire contents of the egg cell.

Wax Layer: The lipid or waxy layer that lies just above the cuticulin layer; it serves as the chief barrier to movement of water into or out of the insect's body.

Basal: Concerning the base of a structure – that part nearest the body. Basal cells in Diptera are generally small cells near the base of the wing.

Basitarsus: The Ist segment of the tarsus–usually the largest.

Batumen: A protective layer of propilis or hard cerumen that encloses the nest cavity of a stingless bee colony.

Benzene Hexachloride: (chemical name) or BHC. (common name). A synthetic insecticide, a chlorinated hydrocarbon, -1,2,3,4,5,6hexachlorocvclohexane of mixed isomers; slightly more toxic to mammals than DDT, acute oral LD51 for rats about 200 mg/kg; phytotoxicity: more toxic than DDT, interferes with germination, suppresses growth and reduces yields except at low concentration; certain crop plants, as potato absorb crude BHC with consequent tainting of tubers.

Bilateral Symmetry: Similarity of form, one side with the other.

Biological Control: The control of pests by employing predators, parasites, or disease; the natural enemies are encouraged and disseminated by man.

Bionomics: The study of the habits, breeding, and adaptations of living forms.

Bipectinate: Feathery, with branches growing out oil both sides of the main axis: applied mainly to antennae.

Bisexual: Having two sexes distinct and separate;i.e. a species with males and females.

Bivouac: The mass of army ant workers within which the queen and brood, live while the colony is not on the move.

Bivoltine: Having two generations per year.

Blastogenesis: The origination of different castes, within a species, from the egg by means other than genetic.

Book Lung: A respiratory cavity containing a series of leaflike folds.

Bot: The larva of certain flies that are parasitic in the body of mammals.

Brachypterous: With short wings that do not cover the abdomen, used of individuals of a species which otherwise has longer wings.

Bract: A small leaf at the base of the flower.

Brood: In insects, a group of individuals of a given species which have hatched into young or which have become adult at approximately the same time and which live together in a defined and limited area. Often referring to the immature stages of ants bees and wasps.

Bubonic Plague: A bacterial disease of rodents and man caused by *Pasteurella pestis* and transmitted chiefly by the oriental rat flea; marked by chills, fever, and inflammatory swelling of lymphatic glands.

Budding: Colony fission, the creation of new colonies by the departure of one or more reproductive females accompanied by a group of workers specifically to establish a new colony.

Bursa Copulatrix: That part of the female genitalia which receives the aedeagus and sperm during copulation. Its structure is often important in separating closely related species.

Caecum: (pl., caeca). A sac or tubelike structure open at only one end.

Calcareous: Referring to soils or rocks, possessing those elements which result in alkaline or basic reactions.

Callow: Newly eclosed workers in social insect colonies whose exoskeletons are still soft and whose colour has not fully matured.

Callus: A rounded swelling: applied especially to swollen regions at the front or back of the thorax in various flies.

Calypter: Innermost of the three flap-like outgrowths at the base of the wing in various flies. Also known as the thoracic squama, it generally conceals the haltere.

Calyptodomous: Of the nests of wasps, referring to those which are surrounded by an envelope.

Campodeiform: (applied to a larva) Grub-like, flattened and elongated with well-developed legs and antennae. Many beetle larvae are of this type, and so are those of the lacewings.

Capitate: With an apical knob like enlargement.

Capitulum: Head like structure of ticks which bears the feeding organs.

Carabiform Larva: A larva shaped like the larva of a carabid beetle, that is etiolate, flattened, and with well-developed legs; with no filaments on the end of the abdomen.

Carbohydrate: Any of a group of neutral compounds made up of carbon, hydrogen, and oxygen; for example, sugar, starch, cellulose.

Cardo: The basal segment of the maxilla or secondary jaw.

Carina: A ridge or keel.

Carnivorous: Preying or feeding on animals.\

Castes: Groups of individuals that become irreversibly behaviorally distinct at some point prior to reproductive maturity. One of three or more distinct forms which make up the population among social insects. The usual three castes are queen, drone (male), and worker. The termites and some of the ants have one or more soldier castes as well.

Caterpillar: The larva of a moth, butterfly, or saw-fly.

Catfacing: The injury caused by the feeding of such insects as plant bugs and stink bugs on developing fruit which results in uneven growth and a deformed mature fruit.

Cauda: The pointed end of the abdomen in aphids.

Caudal: Concerning the tail end.

Cell: An area of the wing bounded by a number of veins. A cell is closed if it is completely surrounded by veins and open if it is bounded partly by the wing margin.

Cellulose: An inert carbohydrate, the chief component of the solid framework or woody part of many plants.

Cement layer: A thin laver on the surface of insect cuticles formed by the hardened secretion of the dermal glands.

Cephalic: Of or pertaining to the head.

Cephalothorax: A body region consisting of head and thoracic segments, as in spiders.

Cerci: (singular: cercus) The paired appendages, often very long, which spring from the tip of the abdomen in many insects.

Cerumen: A mixture of wax and propolis used by social bees in nest construction.

Cervical: Concerning the neck region, just behind the head.

Chaetae: Stiff hairs or bristles (singular: chaeta).

Chaetotaxy: The arrangement of the bristles or chaetae on an insect: especially important in the classification of the Diptera, Collembla and several other groups.

Chelicera:(pl., chelicerae). The anterior pair of appendages in arachnids, the fangs.

Chigger: The parasitic larva of trombiculid mites.

Chitin: The tough horny material, chemically known as a nitrogenous polysaccharide, which makes up the bulk of the insect cuticle, also occurs in other arthropods.

Chorion: The inner shell or covering of the insect egg.

Chromosomes: At cell division the dark-staining, rod-shaped structures which contain the hereditary units called genes.

Chrysalis: The pupa of a butterfly.

Ciliated: Bearing minute hairs (cilia).

Cladogram: A diagram showing nothing more than the sequence in which groups of organisms are interpreted to have originated and diverged in the course of evolution.

Class: A division of the animal kingdom lower than a phylum and higher than an order, for example the class Insecta.

Clavate: Club-shaped, with the distal end swollen: most often applied to antennae.

Claustral Foundation: A way of setting up of a new colony by a queen, or king and queen in the termites, which involves her/them being sealing her/themselves a way in a small chamber and raising the first group of workers entirely (or almost so) on stored body reserves (fat and often the flight muscles).

Clavus: Posterior part of the forewing of of heteropteran bugs.

Cleptoparasitism: Where one female uses the resources and nest of another individual (of either the same or a different species) to provide for her young thus usurping the owners efforts and preventing her from using them.

Cline: A progressive, usually continuous change in one or more characters of a species over a geographic or altitudinal range.

Club: The thickened terminal (farthest from the head) end of the antennae.

Clypeus: Lowest part of the insect face, just above the labrum.

Coarctate: (applied to pupae) Enclosed within the last larval skin, which therefore acts as a cocoon and protects the pupa. Such pupae are found in the flies (Diptera, of the sub-order Cyclorrhapha.).

Cocoon: A case, made partly or completely of silk, which protects the pupa in many insects, especially the moths. The cocoon is made by the larva before it pupates.

Colony: A small or large locally isolated population.

Colony: Of social insects, a group which co-operates in the construction of a nest and in the rearing of the young.

Comb: The grouped cells of the nests of social many hymenoptera.

Comb: A group of spines on the leg of an insect specifically used for cleaning other parts of the insects body.

Commensalism: Symbiosis, one or more individuals from two or more species living together such that one benefits but neither loses fitness.

Commissure: A bridge connecting any two bodies or structures on a body.

Communal: Where females of one species co-operate in nest building but not in brood care.

Complete Metamorphosis or Complex Metamorphosis: Metamorphosis in which the insect develops through four distinct stages, e.g.., ova or egg, larva, pupa, and adult or imago; the wings (when present) develop internally during the larval stage.

Compound eye. An eye consisting of many individual elements or ommatidia each of which is represented externally by a facet.

Connective: A longitudinal cord of nerve fibers connecting successive ganglia.

Contiguous: Touching – usually applied to eyes (see also Holoptic).

Conspecific: Belonging to the same species.

Construction Gland: A gland of wasps producing a size-like substance which enables them to make paper out of wood-pulp.

Copularium: The first chamber built by a newly mated pair of sexual termites.

Corbicula: The pollen basket on the hind leg of many bees, formed by stout hairs on the borders of the tibia.

Corium: The main part of the forewing of a heteropteran bug.

Cornicle: One of the pair of small tubular outgrowths on the hind end of the aphid abdomen.

Corpora Allata: A pair of small endocrine glands located just behind the brain.

Cosmopolitan: Occurring throughout most of the world.

Costa: One of the major longitudinal veins, usually forming the front margin of the wing and usually abbreviated to C. The costal margin is the front edge of the wing.

Costal Cell: The cell between the costa and the sub-costal vein.

Costal Fold: A narrow, thin membrane folded back on the upper surface of the costa of the forewing of butterflies, it contains androconia

Coxa: The basal segment of the insect leg, often immovably attached to the body. **Crawler.** The active first instar of a scale insect.

Cremaster: The cluster of minute hooks (sometimes just one larger hook) at the hind end of a lepidopterous pupa: used to grip the pupal support.

Crochets: (Pronounced crow-shays). Hooked spines at tip of the prolegs of lepidopterous larvae.

Crop: The dilated section of the foregut just behind the esophagus.

Cross-Vein: A short vein joining any two neighboring longitudinal veins.\

Cryptic: Colouring and or pattern adapted for the purpose of protection from predators or prey by concealment.

Cryptobiotic: Leading a hidden or concealed life.

Cubitus: One of the major longitudinal veins, situated in the rear half of the wing and usually with 2 or 3 branches: abbreviated to Cu.

Cuneus: A more or less triangular region of the forewing of certain heteropteran bugs, separated from the corium by a groove or suture.

Cursorial: Adapted for running.

Cuspidal: Two segments of curved lines meeting and terminating at a sharp point.

Cuticle: The outer noncellular layers of the insect integument secreted by the epidermis.

Cyclorrhaphous Diptera: The group of flies which emerge from the puparium through a circular opening at one end of the puparium. These flies belong to the more advanced families.

Cytology: The study of cells and there functioning.

DDT: (common name). A widely used synthetic insecticide; a chlorinated hydrocarbon, dichloro diphenyl trichloroethane.

Dealate: Wingless as a result of the insect casting or breaking off its own wings, as in newly mated queen ants and termites.

Decticous: Of pupa: referring to the state in which the pupa possesses movable mandibles which can be used for biting, the opposite being Adecticous.

Dengue: (pronounced deng'e). A virus disease of man marked by severe pains in head, eyes, muscles, and joints and transmitted by certain mosquitoes.

Dentate: Toothed, possessing teeth or teeth like structures.

Denticulate: Bearing very small tooth-like projections.

Deutonymph: The third instar of a mite.

Diapause: A period of suspended animation of regular occurrence in the lives of many insects, especially in the young stages.

Diaphragm: A horizontal membranous partition of the body cavity.\

Differentiation: Increase in visible distinctive morphology.

Dimorphic: Occurring in two distinct forms.

Dimorphism: A difference in size, form, or color, between individuals of the same species, characterizing two distinct types.\

Discal: The central portion of a wing from the costa to the inner margin.

Discal Cell: Name given to a prominent and often quite large cell near the middle of the wing. The discal cell of one insect group may not be bounded by the same veins as that of another group.

Distad: In a direction away from the body.

Distal: Concerning that part of an appendage furthest from the body.

D.N.A.: An abbreviation for Dioxyribonucleic Acid a large molecule which stores the data in our genes in the form of a 3 character code. D.N.A. is a self replicating molecule.

Dorsal: On or concerning the back or top of an animal.

Dorsal Nectary Organ: In the larvae of many species of Lycaenidae (Blue Butterflies) a gland located in the dorsal region of the 7th abdominal segment, it secretes a sweet substance which is attractive to ants.

Dorsal Ocellus: The simple eye in adult insects and in nymphs and naiads.

Dorsal Shield: The scutum or sclerotized plate covering all or most of the dorsal surface in males and the anterior portion in females, nymphs, and larvae of hard-backed ticks.

Dorso-Central Bristles: The 2 rows of bristles running along the thorax of a fly on the outer side of the acrostichal bristles.

Dorso-Lateral: Towards the sides of the dorsal (upper) surface.

Dorso-Ventral: Running from the dorsal (upper) to the ventral (lower) surface.

Dorsum: The upper surface or back of an animal.

Drone: The male honey bee.

Dulosis: The act of slave making in ants, a species which makes a slave of another is often referred to as Dulotic.

Ecdysis: The moulting process, by which a young insect changes its outer skin or pupal case.

Eclosion: Emergence of the adult or imago from the pupa

Ectoderm: The outer embryological layer which gives rise to the nervous system, integument, and several other parts of an insect.

Ectohormone: A substance secreted by an animal to the outside of its body causing a specific reaction, such as determination of physiological development, in a receiving individual of the same species.

Ectoparasite: A parasite that lives on the outside of its host.

Egg Pod: A capsule which encloses the egg mass of grasshoppers and which is formed through the cementing of soil particles together by secretions of the ovipositing female.

Elateriform Larva: A larva with the form of a wireworm; i.e. long and slender, heavily sclerotized, with short thoracic legs, and with few body hairs.

Elbowed Antenna: Antenna, particularly of ants, in which there is a distinct angle between two of the segments–usually between the 1st and 2nd segments, in which case the 1st segment is usually much longer than the others.

Elytron: (plural elytra) The tough, horny forewing of a beetle or an earwig (See also Hemi-elytron)

Emarginate: With a distinct notch or indentation in the margin.

Emery's Rule: The rule resulting from the observation that species of social parasite are very closely related to their host.

Embolium: A narrow region along the front margin of the forewing in certain heteropteran bugs: separated from the rest of the corium by a groove or suture.

Empodium: An outgrowth between the claws of a fly's foot: it may be bristle-like.

Endemic: Restricted to a well defined geographical region.

Endocrine: Secreting internally, applied to organs whose function is to secrete into blood or lymph a substance which has an important role in metabolism.

Endocuticle: The innermost layer of the cuticle.

Endoparasite: A parasite which lives inside its host's body. Most of the ichneumons, are endoparasites during their larval stages.

Endopterygote: Any insect in which the wings develop inside the body of the early stages and in which there is a complete metamorphosis and pupal stage.

Entomogenous: Growing in or on an insect, for example certain fungi.

Envelope: The carton or wax outermost later of the nest of a social insect, particularly those of wasps.

Enzyme: An organic catalyst, normally a protein formed and secreted by a living cell.

Epicuticle: The thin, non-chitinous, surface layers of the cuticle.

Epidermis: The cellular layer of the integument that secretes or deposits a comparatively thick cuticle on its outer surface.

Epigaeic: Living or foraging primarily above ground, compared to Hypogaeic the opposite.

Epimeron: The posterior part of the side wall of any of the three thoracic segments.

Epinotum: The first abdominal segment when it is fused with the last thoracic one, relating to the higher thin waisted hymenoptera. Also called a propodeum.

Epipharynx: A component of many insect mouth-parts which is attached to the posterior surface of the labrum or clypeus. In chewing insects it is usually only a small lobe, but in the fleas it is greatly enlarged and used for sucking blood.

Epiproct: An appendage arising from the mid-line of the last abdominal segment, just above the anus. In the bristletails and some mayflies it is very long and forms the central 'tail'

Episternum: The anterior part of the side wall of any of the three thoracic segments.

Epithelium: The layer of cells that covers a surface or lines a cavity.

Ergatogyne: Any female member of a eusocial group whose morphological development is somewhere between that of a worker and a queen.

Eruciform: (applied to a larva) Caterpillar like; more or less cylindrical with a well developed head and stumpy legs at the rear as well as the true thoracic legs. The caterpillars of butterflies and moths are typical examples.

Eusocial: A species which lives in a society such that individuals of the species cooperate in caring for the young, which not all of them have produced; there is a reproductive division of labor, with more or less sterile individuals working on behalf of fecund individuals; and there is an overlap of at least two generations in life stages capable of contributing to colony labor, so that offspring assist parents during some period of their life.

Exarate Pupa: A pupa in which all the appendages, legs etc., are free and capable of movement.

Excavate: Hollowed out: applied to the coxae of many beetles, which are hollowed out to receive the femora when the legs are folded.

Excretion: The elimination of the waste products of metabolism.

Exocuticle: The hard and usually darkened layer of the cuticle lying between the endocuticle and epicuticle.

Exoskeleton: Collectively the external plates of the body wall.

Exopterygote: Any insect in which the wings develop gradually on the outside of the body, in which there is only a partial metamorphosis and no pupal stage.

Exuvia: The cast-off outer skin of an insect or other arthropod.

Eye-Cap: Hood formed by the base of the antenna and partly covering the eye in certain small moths.

Fossorial: Adapted for digging.

Foveola: (pl. foveolae) One of the paired depressions on each side of the vertex in grasshoppers.

Frenulum: The wing-coupling mechanism found in many moths.

Frons: Upper part of the insect face, between and below the antennae and usually carrying the median ocellus or simple eye. In true flies (Diptera) it occupies almost all of the front surface of the head apart from the eyes.

Frontal Bristles: The two vertical rows of bristles running down the face of a fly from the ocelli to the antennae

Fronto-Orbital Bristles: The short row of bristles on each side of a fly's head between the eye and the frontal bristles.

Furcula: The forked spring of a springtail.

Girdle: A silken thread supporting the midsection of a pupa.

GIabrous: Without hairs.

Glossa: (plural glossae) One of a pair of lobes at the tip of the labium or lower lip: usually very small, but long in honey bees and bumble bees, in which the two glossae are used to suck up nectar.

Gnathosoma: The anterior part of the body of mites and ticks which bears the mouth and mouthparts.

Gregarious: Living in groups.

Haustellate: Adapted for sucking liquids rather than biting solid food.

Heart: The chambered, pulsatile portion of the dorsal blood vessel.

Head. The anterior body region of insects which bears the mouthparts, eyes, and antennae.

Hematophagous: Feeding or subsisting on blood.

Hemi-elytron: (plural hemi-elytra). The forewing of a heteropteran bug, differing from the beetle elytron in having the distal portion membranous.

Hemimetabola: Insects with simple metamorphosis, with no pupal stage.

Hemimetabolous: Having an incomplete metamorphosis, with no pupal stage in the life history.

Hermaphroditic: Containing the sex organs of both sexes in one individual.

Heteromerous: (of beetles) Having unequal numbers of tarsal segments on the three pairs of legs.

Hexapod: An animal possessing six legs, more specifically the parent group that contains insects and their close kin.

Hibernation: Dormancy during the winter.

Hindgut: The posterior part of the alimentary canal between the midgut and anus.

Insecta: A 'class' of the 'phylum' Arthropoda, distinguished by adults having three body regions: head, thorax, and abdomen; and by having the thorax three-segmented with each segment bearing a pair of legs.

Instar: The stage in an insect's life history between any two moults. A newly-hatched insect which has not yet moulted is said to be a first-instar nymph or larva. The adult (imago) is the final instar.

Integument: The insect's outer coat.

Intermediate Host: The host which harbors the immature stages or the asexual stages of a parasite, a separate organism to that which harbours the sexual stage.

Intercalary Vein: An additional longitudinal vein, arising at the wing margin and running inwards but not directly connected to any of the major veins.

Johnston's organ: A sense organ located in the second antennal segment of many insects and particularly well developed in male mosquitoes and certain other Diptera.

Jugum: A narrow lobe projecting from the base of the forewing in certain moths and overlapping the hind wing, thereby coupling the two wings together.

Keel: A narrow ridge: also called a carina

Labrum-Epipharynx: A mouthpart composed of the labrum and epipharynx and usually elongate.

Lacinia: The inner branch of the maxilla, the outer one being the galea

Lamella: A thin, leaf-like flap or plate, the name being applied to the outgrowths of certain antennae.

Lamellate: Possessing lamellae: applied especially to antennae.

Larva: Name given to a young insect which is markedly different from the adult: caterpillars and fly maggots are good examples.

Maggot: A vermiform larva; a larva without legs and without well-developed head capsule.

Malpighian Tubes: Excretory tubes of insects arising from the anterior end of the hindgut and extending into the body cavity.

Mandible: The jaw of an insect. It may be sharply toothed and used for biting, as in grasshoppers and wasps, or it may be drawn out to form a slender needle as in mosquitoes. Mandibles are completely absent in most flies and lepidopterans.

Mandibulate: Having mandibles suited for biting and chewing.

Marginal Cell: One of a number of cells bordering the front margin of the wing in the outer region.

Maxilla: (plural maxillae) One of the two components of the insect mouth-parts lying just behind the jaws. They assist with the detection and manipulation of food and are often drawn out into tubular structures for sucking up liquids.

Maxillary: Concerning or to do with the maxillae.

Meconium: The reddish fluid ejected by a member of the lepidoptera after emerging from the pupa/chrysalis.

Metatarsus: The basal segment of the tarsus or foot: usually the largest segment.

Metathorax: The 3rd and last segment of the thorax.

Nectar: The sugary liquid secreted by many flowers.

Nectary: A floral gland which secretes nectar.

Neurone: The entire nerve cell including all its processes

Nit: The egg of a louse.

Nocturnal: Active at night.

Nodus.: The kink or notch on the costal margin of the dragonfly wing. The name is also used for the strong, short cross-vein just behind the notch.

Notaulix: One of a pair of longitudinal grooves on the mesonotum of certain hymenopterans, dividing the mesonotum into a central area and two lateral areas (plural notaulices)

Notopleuron: A triangular area on the thorax of certain flies, just behind the humeral callus and occupying parts of both dorsal and lateral surfaces.

Notum: The dorsal or upper surface of any thoracic segment: usually prefixed by pro-, meso-, or meta- to indicate the relevant segment.

Nucleus: The spheroid body within a cell that has the major role in controlling and regulating the cell's activities and contains the hereditary units or genes.

Nurse Cells: Cells that are located in the ovarian tubes of certain insects and that furnish nutriment to the developing eggs.

Nymph: Name given to the young stages of those insects which undergo a partial metamorphosis. The nymph is usually quite similar to the adult except that its wings are not fully developed. It normally feeds on the same kind of food as the adult.

Occipital Suture: A groove running round the posterior region of the head of some insects and separating the vertex from the occiput. On the sides of the head the same groove marks the posterior boundary of the cheeks or genae

Occiput: Hindmost region of the top of the head, just in front of the neck membrane. In some insects it is separated from the vertex by the occipital suture, but it is not usually present as a distinct plate or sclerite.

Ocellar Bristles: Bristles arising around or between the ocelli in various flies.

Ocellar Triangle: A triangular area, usually quite distinct from the rest of the head, on which the ocelli of true flies are carried.

Ocellus: (Plural Ocelli) One of the simple eyes of insects, usually occurring in a group of three on the top of the head, although one or more may be absent from many insects.

Oesophagus: The narrow part of the alimentary canal immediately posterior to the pharynx and mouth.

Ommatidium: (pl., ommatidia). One of the units which make up the compound eyes of arthropods.

Ootheca: (pl., oothecae). An egg case formed by the secretions of accessory genital glands or oviducts, such as the purse-like structure carried around by cockroaches or the spongy mass in which mantids lay their eggs

Oral Vibrissae: The pair of large bristles just above the mouth in certain flies: usually simply called vibrissae.

Order: A subdivision of a class or subclass containing a group of related families. Organophosphates. Organic compounds containing phosphorous; an important group of synthetic insecticides belong to this class of chemicals.

Oribatid Mite: A mite belonging to the Oribatei, a large unit of mites containing about 35 families in the suborder Sarcoptiformes.

Oviparous: Producing eggs which are hatched outside the body of the female.

Ovipositor: The tubular or valved egg-laying apparatus of a female insect: concealed in many insects, but extremely large among the bush-crickets and some parasitic hymenopterans.

Ovoviviparous: Producing living young by the hatching of the egg while still within the female.

Pharynx: The anterior part of the foregut between the mouth and the esophagus.

Pheromone: A substance secreted by an animal which when released externally in small amounts causes a specific reaction, such as stimulation to mate with or supply food to a receiving individual of the same species.

Phoresis: The usage by one animal of another soley as a means of transport, i.e. certain mites on various other insects.

Phylu: (pl., phyla). A major division of the animal kingdom, containing various suborders and classes etc.

Phytophagous: Feeding upon plants.

Phytotoxic: Poisonous to plants.

Platyform larva: A very flattened larva.

Plumose: Feather-like, as in plumose antennae

Pictured: A term used to describe wings, especially among the Diptera, which have dark mottling on them.

Pilose: Densely clothed with hair.

Pleural: Concerning the side walls of the body.

Pleural Suture: A vertical or diagonal groove on each of the thoracic pleura, separating the episternum at the front from the epimeron at the back.

Pleuron.: The side wall of a thoracic segment.

Plumose: With numerous feathery branches: applied especially to antennae.

Predator: An animal that attacks and feeds on other animals, usually smaller and weaker than itself.

Prementum: The distal region of the labium, from which spring the labial palps and the ligula.

Preovipositional Period: The period between the emergence of an adult female and the start of its egg laying.

Prepupa: The last larval instar after it ceases to feed; often it takes on a distinctive appearance becoming quiescent and rather shrunken, and often looks dead.

Prothorax: The 1st or anterior thoracic segment.

Protonymph: The second instar of a mite.

Proventriculus: The posterior section of the foregut.

Pseudoscorpions: Small arachnids, seldom over 5 mm. long, scorpion-like in general appearance but without sting.

Pseudovipositor: The slender tube to which the posterior part of the abdomen is reduced in the female of certain insects.

Proximal: Concerning the basal part of an appendage – the part nearest to the body.

Quadrilateral: A cell near the base of the damselfly wing, whose shape is important in separating the families.

Queen cell: The special cell in which a queen honey bee de Reticulate. Covered with a network pattern.

Reproductives: In termites the caste of kings and queens in other social insects only the queens.

Rostrum: A beak or snout, applied especially to the piercing mouth-parts of bugs and the elongated snouts of weevils.

Rudimentary: Poorly or imperfectly developed.

Saprophytic: Living on dead or decaying organic matter.

Scale: A scale insect; a member of the order Homoptera.

Sebaceous Gland: A gland producing a greasy lubricating substance.

Secondary Parasite: A parasite on another parasite.

Segment: One of the rings or divisions of the body, or one of the sections of a jointed limb.

Segmentation: The embryological process by which the insect body becomes divided into a series of parts or segments.

Serrate: Toothed like a saw.

Suture: A groove on the body surface which usually divides one plate or sclerite from the next: also the junction between the elytra of a beetle.

Tarsus: (pl., tarsi). The insect's foot: primitively a single segment but consisting of several segments in most living insects.

Tegmen: (plural tegmina) The leathery forewing of a grasshopper or similar insect, such as a cockroach

Tegula: A small lobe or scale overlying the base of the forewing like a shoulder-pad.

Tergite: The primary plate or sclerite forming the dorsal surface of anybody segment.

Truncate: Ending abruptly: squared off.

Tubercle: A small knob like or rounded protuberance.

Tymbal: The sound-producing 'drum-skin' of a cicada.

Tympanum: The auditory membrane or ear-drum of various insects.

Uric Acid: The chief nitrogenous waste of birds, reptiles and insects-; chemically, C, H, N, O,.

Veins: In insects, the rib like tubes that strengthen the wings.

Vermiform larva: A legless wormlike larva without a well-developed head

Venation: The arrangement of veins in the wings of insects.Ventral. Concerning the lower side of the body.

Vertex: The top of the head, between and behind the eyes.

Vestigial: Poorly developed, degenerate or atrophied, more fully functional in an earlier stage of development of the individual or species.

Visceral Muscle: A muscle which invests an internal organ.

Vibrissae: The pair of large bristles just above the mouth in certain flies: usually simply called vibrissae.

Wing Pads: The undeveloped wings of nymphs and naiads, which appear as two flat structures on each side.

Woollybear: A very hairy caterpillar belonging to the family Arctiidae, the tiger moths.

REFERENCES

Metcalf C.L. and Flint W.P., Destructive and Useful Insects. McGraw-Hill Co., New York., (9), 67 (1962)

Hilton Pond Center, The importance of pollinators, http://www.hiltonpond.org/ This Week 031008.html (2003)

Pickett C.H. and Bugg R.L., Enhancing biological control: habitat management to promote natural enemies of agricultural pests, Berkeley, CA: University of California Press, 421 (1998)

Ross H. H., Ross C.A. and Ross J. R., A Textbook of Entomology, John Wiley and Sons, New York, NY, figure-3, 611, figure-7, 609, figure-11, 618 (1982)

Eraldo Medeiros and Costa. Neto., Entomotherapy, or the Medicinal Use of Insects, J. of Ethnobiol., 25(1), 93-114 (2005)

Cicero K., Making a home for beneficial insects, The New Farm, February, 28-33 (1993)

Colley M.R. and Luna J.M., Relative attractiveness of potential beneficial insectary plants to aphidophagous hoverflies, Environ. Entomol., (29), 1054-1059 (2000)

Jones G.A. and Gillett J.L., Intercropping with sunflowers to attract beneficial insects in organic agriculture, Florida Entomologist., (88), 91-96 (2005

Alam, M.A. (2010). Encyclopedia of Applied Entomology. Anmol Publications Pvt. Ltd., New Delhi, India.

Buurma, J.S. (2008). Stakeholder involvement in crop protection policy planning in the Netherlands. ENDURE – RA3.5/SA4.5 Working Paper. LEI Wageningen UR, The Hague, The Netherland.

Dhaliwal, G.S. and O. Koul (2007). Biopesticides and Pest Management: Conventional and Biotechnological Approaches. Kalyani Publishers, Ludhiana, New Delhi, India.

Dhaliwal, G.S. and R. Arora (2003). Principles of Insect Pest Management, 2nd edition, Kalyani Publishers, Ludhiana, India, 90-94.

Dhaliwal, G.S., R. Singh and B.S. Chhillar (2006). Essentials of Agricultural Entomology. Kalyani Publishers, Ludhiana, New Delhi, India.

Elton, C. (1939). Animal Ecology. Macmillan, New York, USA. Farrell, K.R. (1990). IPM: Reshaping the approach to pest management. Cal. Agri. 44: 2-15.

Flint, M.L. and R. Bosch (1981). Introduction to Integrated Pest Management. Plenum Press, New York, USA. Flint, M.L., S. Daar and R. Molinar (2003). Establishing Integrated Pest Management Aolicies and Programs: A guide for public agencies. University of California, Division of Agriculture and Natural Resources, Communication Services, Oakland, California, USA.

Food and Agriculture Organization (FAO) (2014). International Code of Conduct on the Distribution and Use of Pesticides: Guidelines on Prevention and Management of Pesticide Resistance. Chief, Publishing Policy and Support Branch, Office of Knowledge Exchange, Research and Extension, FAO, Viale delle Terme di Caracalla, 00153 Rome, Italy.

Grant, W.P., J. Greaves, D. Chandler, G. Davidson and G.M. Tatchell (2003). Design principles for a better regulatory system for biopesticides. University of Warwick, Coventry CV4 7AL, UK. https://www.google.com.pk/Design+principles+for+a+better+regulatory+ system+for+biopesticides&oq/PXjiOxxg. Accessed on 15 th March, 2017.

Hoffmann, J.A., F.C. Kafatos, C.A. Janeway, R.A. Ezekowitz (1999). Phylogenetic perspectives in innate immunity. Sci. 284: 1313–1318.

Hoffmann, M.P. and A.C. Frodsham (1993). Natural enemies of vegetable insect pests. Cooperative Extension, Cornell University, Ithaca, NY. USA.

Inayatullah, C. (1995). Training Manual: Integrated Insect Pest Management. Entomological Research Laboratories and NARC Training Institute, Islamabad.

Jha, L.K. (2010). Applied Agricultural Entomology. New Central Book Agency (Pvt.) Ltd., Kolkata Pune, Delhi, India. Joshi, S.R. (2006). Biopesticides: A Biotechnological Approach. New Age International (Pvt.) Ltd. Publishers, New Delhi, India.

Juang, J.Y., G.G. Katul, A. Porporato, P.C. Stoy, M.S. Sequeira, M. Detto, H.S. Kim, and R. Oren (2007). Eco-hydrological controls on summertime convective rainfall triggers. Global Change Biol. 13:887-896.

Knipling (1979). The Basic Principles of Insect Population Suppression and Management. US Government Printing Office, Department of Agriculture, Washington, DC, USA, Agriculture Handbook No. 512.

Kogan, M. (1988). Integrated pest management theory and practice. Entomol. Exp. Appl. 49: 59-70.

Lewis, V.R. and S.J. Seybold (2010). Wood-boring beetles in homes. Pest Notes Publication 7418, University of California Statewide Integrated Pest Management Program Agriculture and Natural Resources, pp. 1-4.

Matthews, G.A. (2000). Pesticides Application Methods. 3rd Edition, Blackwell Science Ltd., USA.

Metcalf, R.L. and W.H. Luckmann (1975). The pest management concept. In: Metcalf, R.L. and W.H. Luckmann (eds). Introduction to Insect Pest Management. New York: John Wiley and Sons. pp. 3–35.

Nasrin, S. (2016). Insect pest of herbarium and their integrated pest management. Int. J. Human. Soc. Stud. 4: 68-71.

Norris, R.F., E.P. Caswell-Chen and M. Kogan (2002). Concepts in Integrated Pest Management. Prentice-Hall of India Private Limited, New Delhi, India.

O'Connor-Marer, P. (2006). Residential, Industrial, and Institutional Pest Control. Oakland: Univ. Calif. Agric. Nat. Res. Publ. 3334.

Omafra, S. (2009). Agronomy Guide for Field Crops: Field Scouting. Ministry of Agriculture and Food, Ontario. http://www.omafra.gov.on.ca/english/crops/pub811/10 field.htm. Accessed on 20 August, 2016.

Pattison, M. (2005). The Quiet Revolution: Push–Pull Technology and the African Farmer. Gatsby Charitable Foundation, London, UK.

Pedigo, L.P. and M.E. Rice (2009). Entomology and Pest Management. 6th edition, PHI Learning Private Limited, New Delhi, India.

Peterson, C., T.L. Wagner, J.E. Mulrooney and T.G. Shelton (2006). Subterranean Termite, Their Prevention and Control in Building. U.S. Department of Agriculture (USDA) Forest Service Wood Products Insect Research Unit, 201 Lincoln Green, Starkville, MS 39759.

Rondon, S.I., G.H. Clough and M.K. Corp (2008). How to identify, scout, and control insect pests in vegetable crops. Department of Agriculture, and Oregon counties, Oregon State University, U.S.A. http://hdl.handle.net/1957/19902. Accessed on 02 February 2015.

Saha, L.R. and G.S. Dhaliwal (2012). Handbook of Plant Protection, 2nd Edition, Kalyani Publishers, New Delhi, India Schillhorn, V.V.T., D. Forno, S. Joffe, D. Umali-Deininger and S. Cooke (1997). Integrated Pest Management: Strategies and Policies for Effective Implementation. Environmentally Sustainable Development Studies and Monographs, No. 13. World Bank, Washington, DC, USA

Schowalter, T.D. (2011). Insect Ecology: An Ecosystem Approach. Academic Press London NW1 7BY, UK. Singh, R. (2008). Crop Protection by Botanical Pesticides. CBS Publishers and Distributors, New Delhi, India.

Sorby, K., G. Fleischer and E. Pehu (2005). Integrated pest management in development: review of trends and implementation strategies. Agriculture and Rural Development Working Paper 5, World Bank, Washington, DC. http://documents.worldbank. org/curated/en/ 2003/04/2455449/integrated-pestmanagement-development-eviewtrendsimplementation- strategies. Accessed on 25 December, 2016. SP-IPM, (2008). Incorporating Integrated Pest Management into National Policies. IPM Research Brief No. 6. SP-IPM Secretariat, International Institute of Tropical Agriculture (IITA), Ibadan, Nigeria. www.spipm.cgiar.org. Accessed on 08th July, 2015.

Trenberth, K.E. (1999). Atmospheric moisture recycling: role of advection and local evaporation. J. Clim. 12: 1368-1381.

Trivedi, P.C. (2002). Plant Pest Management. Aavishkar Publishers Distributors, Diamond Publishing Press, Jaipur, India.

Van-der-Wulp, H. and J. Pretty (2005). Policies and trends. In: Pretty, J.N. (ed). The Pesticide Detox: Towards a More Sustainable Agriculture, 1st Edition, Earthscan, London, UK. 226-249.

INDEX